T0329656

Model-Based Reinforcement Learning

Model-Based Reinforcement Learning

From Data to Continuous Actions with a Python-based Toolbox

Milad Farsi and Jun Liu
University of Waterloo, Ontario, Canada

IEEE Press Series on Control Systems Theory and Applications
Maria Domenica Di Benedetto, Series Editor

WILEY

Published by John Wiley & Sons, Inc., Hoboken, New Jersey.
Published simultaneously in Canada.

For general information on our other products and services or for technical support, please contact our Customer Care Department within the United States at (800) 762-2974, outside the United States at (317) 572-3993 or fax (317) 572-4002.

Wiley also publishes its books in a variety of electronic formats. Some content that appears in print may not be available in electronic formats. For more information about Wiley products, visit our web site at www.wiley.com.

Library of Congress Cataloging-in-Publication Data applied for:

Hardback ISBN: 9781119808572

Cover Design: Wiley
Cover Images: © Pobytov/Getty Images; Login/Shutterstock; Sazhnieva Oksana/Shutterstock

Set in 9.5/12.5pt STIXTwoText by Straive, Chennai, India

Contents

About the Authors

Milad Farsi received a B.S. degree in Electrical Engineering (Electronics) from the University of Tabriz in 2010. He obtained an M.S. degree also in Electrical Engineering (Control Systems) from Sahand University of Technology in 2013. Moreover, he gained industrial experience as a Control System Engineer between 2012 and 2016. Later, he acquired a Ph.D. degree in Applied Mathematics from the University of Waterloo, Canada, in 2022 and is currently a Postdoctoral Fellow at the same institution. His research interests include control systems, reinforcement learning, and their applications in robotics and power electronics.

Jun Liu received a B.S. degree in Applied Mathematics from Shanghai Jiao-Tong University in 2002, the M.S. degree in Mathematics from Peking University in 2005, and the Ph.D. degree in Applied Mathematics from the University of Waterloo, Canada, in 2010. He is currently an Associate Professor of Applied Mathematics and a Canada Research Chair in Hybrid Systems and Control at the University of Waterloo, where he directs the Hybrid Systems Laboratory. From 2012 to 2015, he was a Lecturer in Control and Systems Engineering at the University of Sheffield. During 2011 and 2012, he was a Postdoctoral Scholar in Control and Dynamical Systems at the California Institute of Technology. His main research interests are in the theory and applications of hybrid systems and control, including rigorous computational methods for control design with applications in cyber-physical systems and robotics.

Preface

The subject of Reinforcement Learning (RL) is popularly associated with the psychology of animal learning through a trial-and-error mechanism. The underlying mathematical principle of RL techniques, however, is undeniably the theory of optimal control, as exemplified by landmark results in the late 1950s on dynamic programming by Bellman, the maximum principle by Pontryagin, and the Linear Quadratic Regulator (LQR) by Kalman. Optimal control itself has its roots in the much older subject of calculus of variations, which dated back to late 1600s. Pontryagin's maximum principle and the Hamilton–Jacobi–Bellman (HJB) equation are the two main pillars of optimal control, the latter of which provides feedback control strategies through an optimal value function, whereas the former characterizes open-loop control signals.

Reinforcement learning was developed by Barto and Sutton in the 1980s, inspired by animal learning and behavioral psychology. The subject has experienced a resurgence of interest in both academia and industry over the past decade, among the new explosive wave of AI and machine learning research. A notable recent success of RL was in tackling the otherwise seemingly intractable game of Go and defeating the world champion in 2016.

Arguably, the problems originally solved by RL techniques are mostly discrete in nature. For example, navigating mazes and playing video games, where both the states and actions are discrete (finite), or simple control tasks such as pole balancing with impulsive forces, where the actions (controls) are chosen to be discrete. More recently, researchers started to investigate RL methods for problems with both continuous state and action spaces. On the other hand, classical optimal control problems by definition have continuous state and control variables. It seems natural to simply formulate optimal control problems in a more general way and develop RL techniques to solve them. Nonetheless, there are two main challenges in solving such optimal control problems from a computational perspective. First, most techniques require exact or at least approximate model information. Second, the computation of optimal value functions and feedback controls often suffers from the curse of dimensionality. As a result, such methods are often too slow to be applied in an online fashion.

The book was motivated by this very challenge of developing computationally efficient methods for online learning of feedback controllers for continuous control problems. A main part of this book was based on the PhD thesis of the first author, which presented a Structured Online Learning (SOL) framework for computing feedback controllers by forward integration of a state-dependent differential Riccati equation along state trajectories. In the special case of Linear Time-Invariant (LTI) systems, this reduces to solving the well-known LQR problem without prior knowledge of the model. The first part of the book (Chapters 1–3) provides some background materials including Lyapunov stability analysis, optimal control, and RL for continuous control problems. The remaining part (Chapters 4–9) discusses the SOL framework in detail, covering both regulation and tracking problems, their further extensions, and various case studies.

The first author would like to convey his heartfelt thanks to those who encouraged and supported him during his research. The second author is grateful to the mentors, students, colleagues, and collaborators who have supported him throughout his career. We gratefully acknowledge financial support for the research through the Natural Sciences and Engineering Research Council of Canada, the Canada Research Chairs Program, and the Ontario Early Researcher Award Program.

Waterloo, Ontario, Canada   *Milad Farsi and Jun Liu*
April 2022

Acronyms

ACCPM	analytic center cutting-plane method
ARE	algebraic Riccati equation
DNN	deep neural network
DP	dynamic programming
DRE	differential Riccati equation
FPRE	forward-propagating Riccati equation
GD	gradient descent
GUAS	globally uniformly asymptotically stable
GUES	globally uniformly exponentially stable
HJB	Hamilton–Jacobi–Bellman
KDE	kernel density estimation
LMS	least mean squares
LQR	linear quadratic regulator
LQT	linear quadratic tracking
LS	least square
LTI	linear time-invariant
MBRL	model-based reinforcement learning
MDP	Markov decision process
MIQP	mixed-integer quadratic program
MPC	model predictive control
MPP	maximum power point
MPPT	maximum power point tracking
NN	neural network
ODE	ordinary differential equation
PDE	partial differential equation
PE	persistence of excitation
PI	policy iteration
PV	Photovoltaic
PWA	piecewise affine

PWM	pulse-width modulation
RL	reinforcement learning
RLS	recursive least squares
RMSE	root mean square error
ROA	region of attraction
SDRE	state-dependent Riccati equations
SINDy	sparse identification of nonlinear dynamics
SMC	sliding mode control
SOL	structured online learning
SOS	sum of squares
TD	temporal difference
UAS	uniformly asymptotically stable
UES	uniformly exponentially stable
VI	value iteration

Introduction

I.1 Background and Motivation

I.1.1 Lack of an Efficient General Nonlinear Optimal Control Technique

Optimal control theory plays an important role in designing effective control systems. For linear systems, a class of optimal control problems are solved successfully under the framework of Linear Quadratic Regulator (LQR). LQR problems are concerned with minimizing a quadratic cost for linear systems in terms of the control input and state, solving which allows us to regulate the state and the control input of the system. In control systems applications, this provides an opportunity to specifically regulate the behavior of the system by adjusting the weighting coefficients used in the cost functional. However, when it turns to nonlinear dynamical systems, there is no systematic method for efficiently obtaining an optimal feedback control for the general nonlinear systems. Thus, many of the techniques available in the literature on linear systems do not apply in general.

Despite the complexity of nonlinear dynamical systems, they have attracted much attention from researchers in recent years. This is mostly because of their practical benefits in establishing a wide variety of applications in engineering, including power electronics, flight control, and robotics, among many others. Considering the control of a general nonlinear dynamical system, optimal control involves finding a control input that minimizes a cost functional that depends on the controlled state trajectory and the control input. While such a problem formulation can cover a wide range of applications, how to efficiently solve such problems remains a topic of active research.

I.1.2 Importance of an Optimal Feedback Control

In general, there exist two well-known approaches to solving such optimal control problems: the maximum (or minimum) principles [Pontryagin, 1987] and the Dynamic Programming (DP) method [Bellman and Dreyfus, 1962]. To solve an optimization problem that involves dynamics, maximum principles require us to solve a two-point boundary value problem, where the solution is not in a feedback form.

There exist plenty of numerical techniques presented in the literature to solve the optimal control problem. Such approaches generally rely on knowledge of the exact model of the system. In the case where such a model exists, the optimal control input is obtained in the open-loop form as a time-dependent signal. Consequently, implementing these approaches in real-world problems often involves many complications that are well known by the control community. This is because of the model mismatch, noises, and disturbances that greatly affect the online solution, causing it to diverge from the preplanned offline solution. Therefore, obtaining a closed-loop solution for the optimal control problem is often preferred in such applications.

The DP approach analytically results in a feedback control for linear systems with a quadratic cost. Moreover, employing the Hamilton-Jacobi-Bellman (HJB) equation with a value function, one might manage to derive an optimal feedback control rule for some real-world applications, provided that the value function can be updated in an efficient manner. This motivates us to consider conditions leading to an optimal feedback control rule that can be efficiently implemented in real-world problems.

I.1.3 Limits of Optimal Feedback Control Techniques

Consider an optimal control problem over an infinite horizon involving a non-quadratic performance measure. Using the idea of inverse optimal control, the cost functional can be then be evaluated in closed form as long as the running cost depends somehow on an underlying Lyapunov function by which the asymptotic stability of the nonlinear closed-loop system is guaranteed. Then it can be obtained that the Lyapunov function is indeed the solution of the steady-state HJB equation. Although such a formulation allows analytically obtaining an optimal feedback rule, choosing the proper performance measure may not be trivial. Moreover, from a practical point of view, because of the nonlinearity in the performance measure, it might cause unpredictable behavior.

A well-studied method for solving an optimal control problem online is employing a value function assuming a given policy. Then, for any state, the value function gives a measure of how good the state is by collecting the cost starting from that state while the policy is applied. If such a value function can be obtained, and the system model is known, the optimal policy is actually the one that takes the system in the direction by which the value decreases the most in the space of the states. Such Reinforcement Learning (RL) techniques, which are known as value-based methods, including the Value Iteration (VI) and the Policy Iteration (PI) algorithms, are shown to be effective in finite state and control spaces. However, the computations cannot efficiently scale with the size of the state and control spaces.

I.1.4 Complexity of Approximate DP Algorithms

One way of facilitating the computations regarding the value updates is employing an approximate scheme. This is done by parameterizing the value function and adjusting the parameters in the training process. Then, the optimal policy given by the value function is also parameterized and approximated accordingly. The complexity of any value update depends directly on the number of parameters employed, where one may try limiting the number of the parameters by sacrificing the optimality. Therefore, we are motivated to obtain a more efficient update rule for the value parameters, rather than limiting the number of the parameters. We achieve this by reformulating the problem with a quadratically parameterized value function.

Moreover, the classical VI algorithm does not explicitly use the system model for evaluating the policy. This benefits applications in that the full knowledge of the system dynamics is no longer required. However, online training with VI alone may take much longer time to converge, since the model only participates implicitly through the future state. Therefore, the learning process can be potentially accelerated by introducing the system model. Furthermore, this creates an opportunity for running a separate identifier unit, where the model obtained can be simulated offline to complete the training or can be used for learning optimal policies for different objectives.

It can be shown that the VI algorithm for linear systems results in a Lyapunov recursion in the policy evaluation step. Such a Lyapunov equation in terms of the system matrices can be efficiently solved. However, for the general nonlinear case, methods for obtaining an equivalent are not amenable to efficient solutions. Hence, we are motivated to investigate the possibility of acquiring an efficient update rule for nonlinear systems.

I.1.5 Importance of Learning-based Tracking Approaches

One of the most common problems in the control of dynamical systems is to track a desired reference trajectory, which is found in a variety of real-world applications. However, designing an efficient tracking controller using conventional methods often necessitates a thorough understanding of the model, as well as computations and considerations for each application. RL approaches, on the other hand, propose a more flexible framework that requires less information about the system dynamics. While this may create additional problems, such as safety or computing limits, there are already effective outcomes from the use of such approaches in real-world situations. Similar to regulation problems, the applications of tracking control can benefit from Model-based Reinforcement Learning (MBRL) that can handle the parameter updates more efficiently.

I.1.6 Opportunities for Obtaining a Real-time Control

In the approximate optimal control technique, employing a limited number of parameters can only yield a local approximation of the model and the value function. However, if an approximation within a larger domain is intended, a considerably higher number of parameters may be needed. As a result, the identification and the controller's complexity might be rather too high to be performed online in real-world applications. This convinces us to circumvent this constraint by considering a set of local simple learners instead, in a piecewise approach.

As mentioned, there exist already interesting real-world applications of MBRL. Motivated by this, in this monograph, we aim on introducing automated ways of solving optimal control problems that can replace the conventional controllers. Hence, detailed applications of the proposed approaches are included, which are demonstrated with numerical simulations.

I.1.7 Summary

The main motivation for this monograph can be summarized as follows:

- Optimal control is highly favored, while there is no general analytical technique applicable to all nonlinear systems.
- Feedback control techniques are known to be more robust and computationally efficient compared to the numerical techniques, especially in the continuous space.
- The chance of obtaining a feedback control in closed form is low, and the known techniques are limited to some special classes of systems.

- Approximate DP provides a systematic way of obtaining an optimal feedback control, while the complexity grows significantly with the number of parameters.
- An efficient parameterization of the optimal value may provide an opportunity for more complex real-time applications in control regulation and tracking problems.

I.1.8 Outline of the Book

We summarize the main contents of the book as follows:

- Chapter 1 introduces Lyapunov stability analysis of nonlinear systems, which are used in subsequent chapters for analyzing the closed-loop performance of the feedback controllers.
- Chapter 2 formulates the optimal control problem and introduces the basic concepts of using the HJB equation to characterize optimal feedback controllers, where LQR is treated as a special case. A focus is on optimal feedback controllers for asymptotic stabilization tasks with an infinite-horizon performance criterion.
- Chapter 3 discusses PI as a prominent RL technique for solving continuous optimal control problems. PI algorithms for both linear and nonlinear systems with and without any knowledge of the system model are discussed. Proofs of convergence and stability analysis are provided in a self-contained manner.
- Chapter 4 presents different techniques for learning a dynamic model for continuous control in terms of a set of basis functions, including least squares, recursive least squares, gradient descent, and sparse identification techniques for parameter updates. Comparison results are shown using numerical examples.
- Chapter 5 introduces the Structured Online Learning (SOL) framework for control, including the algorithm and local analysis of stability and optimality. The focus is on regulation problems.
- Chapter 6 extends the SOL framework to tracking with unknown dynamics. Simulation results are given to show the effectiveness of the SOL approach. Numerical results on comparison with alternative RL approaches are also shown.
- Chapter 7 presents a piecewise learning framework work as a further extension of the SOL approach, where we limit to linear bases, while allowing models to be learned in a piecewise fashion. Accordingly, closed-loop stability guarantees are provided with Lyapunov analysis facilitated by Mixed-Integer Quadratic Program (MIQP)-based verification.

- Chapters 8 and 9 present two case studies on Photovoltaic (PV) and quadrotor systems. Chapter 10 introduces the associated Python-based tool for SOL.

It should be noted that, some of the contents of chapters 5–9 have been previously published in Farsi and Liu [2020, 2021], Farsi et al. [2022], Farsi and Liu [2022b, 2019], and they are included in this book with the permission of the cited publishers.

I.2 Literature Review

I.2.1 Reinforcement Learning

RL is a well-known class of machine learning methods that are concerned with learning to achieve a particular task through interactions with the environment. The task is often defined by some reward mechanism. The intelligent agent has to take actions in different situations. Then, the reward accumulated is used as a measure to improve the agent's actions in future, where the objective is to accumulate as much as rewards as possible over some time. Therefore, it is expected that the agent's actions approach the optimal behavior in a long term. RL has gained a lot of successes in the simulation environment. However, the lack of explainability [Dulac-Arnold et al., 2019] and data efficiency [Duan et al., 2016] make them less favorable as an online learning technique that can be directly employed in real-world problems, unless there exists a way to safely transfer the experience from simulation-based learning to the real world. The main challenges in the implementations of the RL techniques are discussed in Dulac-Arnold et al. [2019]. Numerous studies are done on this subject; see, e.g. Sutton and Barto [2018], Wiering and Van Otterlo [2012], Kaelbling et al. [1996], and Arulkumaran et al. [2017] for a list of related works. RL has found a variety of interesting applications in robotics [Kober et al., 2013], multiagent systems [Zhang et al., 2021; Da Silva and Costa, 2019; Hernandez-Leal et al., 2019], power systems [Zhang et al., 2019; Yang et al., 2020], autonomous driving [Kiran et al., 2021] and intelligent transportation [Haydari and Yilmaz, 2020], and healthcare [Yu et al., 2021], among others.

I.2.2 Model-based Reinforcement Learning

MBRL techniques, as opposed to model-free methods in learning, are known to be more data efficient. Direct model-free methods usually require enormous data and hours of training even for simple applications [Duan et al., 2016], while model-based techniques can show optimal behavior in a limited number of trials. This property, in addition to the flexibilities in changing learning objectives and performing further safety analysis, makes them more suitable for real-world implementations, such as robotics [Polydoros and Nalpantidis, 2017]. In model-based approaches, having a deterministic or probabilistic description of the transition system saves much of the effort spent by direct methods in treating any point in the state-control space individually. Hence, the role of model-based techniques becomes even more significant when it comes to problems with continuous controls rather than discrete actions [Sutton, 1990; Atkeson and Santamaria, 1997; Powell, 2004].

In Moerland et al. [2020], the authors provide a survey of some recent MBRL methods which are formulated based on Markov Decision Processes (MDPs). In general, there exist two approaches for approximating a system: parametric and nonparametric. Parametric models are usually preferred over nonparametric, since the number of the parameters is independent of the number of samples. Therefore, they can be implemented more efficiently on complex systems, where many samples are needed. On the other hand, in nonparametric approaches, the prediction for a given sample is obtained by comparing it with a set of samples already stored, which represent the model. Therefore, the complexity increases with the size of the dataset. In this book, because of this advantage of parametric models, we focus on the parametric techniques.

I.2.3 Optimal Control

Let us specifically consider implementations of RL on control systems. Regardless of the fact that RL techniques do not require the dynamical model to solve the problem, they are in fact intended to find a solution for the optimal control problem. This problem is extensively investigated by the control community. The LQR problem has been solved satisfactorily for linear systems using Riccati equations [Kalman, 1960], which also ensure system stability for infinite-horizon problems However, in the case of nonlinear systems, obtaining such a solution is not trivial and requires us to solve the HJB equation, either analytically or numerically, which is a challenging task, especially when we do not have knowledge of the system model.

Model Predictive Control (MPC) [Camacho and Alba, 2013; Garcia et al., 1989; Qin and Badgwell, 2003; Grüne and Pannek, 2017; Garcia et al., 1989; Mayne and Michalska, 1988; Morari and Lee, 1999] has been frequently used as an optimal control technique, which is inherently model-based. Furthermore, it deals with the control problem only across a restricted prediction horizon. For this reason, and for the fact that the problem is not considered in the closed-loop form, stability analysis is hard to establish. For the same reasons, the online computational complexity is considerably high, compared to a feedback control rule that can be efficiently implemented.

Forward-Propagating Riccati Equation (FPRE) [Weiss et al., 2012; Prach et al., 2015] is one of the techniques presented for solving the LQR problem. Normally, the Differential Riccati Equation (DRE) is solved backward from a final condition. In an analogous technique, it can be solved in forward time with some initial condition instead. A comparison between these two schemes is given in Prach et al. [2015]. Employing forward-integration methods makes it suitable for solving the problem for time-varying systems [Weiss et al., 2012; Chen and Kao, 1997] or in the RL setting [Lewis et al., 2012], since the future dynamics are not needed, whereas the backward technique requires the knowledge of the future dynamics from the final condition. FPRE has been shown to be an efficient technique for finding a suboptimal solution for linear systems, while, for nonlinear systems, the assumption is that the system is linearized along the system's trajectories.

State-dependent Riccati Equations (SDRE) [Çimen, 2008; Erdem and Alleyne, 2004; Cloutier, 1997] is another technique that can be found in the literature for solving the optimal control problem for nonlinear systems. This technique relies on the fact that any nonlinear system can be written in the form of a linear system with state-dependent matrices. However, this conversion is not unique. Hence, a suboptimal solution is expected. Similar to MPC, it does not yield a feedback control rule since the control at each state is computed by solving a DRE that depends on the system's trajectory.

I.2.4 Dynamic Programming

Other model-based approaches can be found in the literature that are mainly categorized under RL in two groups: value function and policy search methods. In value function-based methods, known also as approximate/adaptive DP techniques [Wang et al., 2009; Lewis and Vrabie, 2009; Balakrishnan et al., 2008], a value function is used to construct the policy. On the other hand, policy search methods directly improve the policy to achieve optimality. Adaptive DP has found different applications [Prokhorov, 2008; Ferrari-Trecate et al., 2003; Prokhorov et al., 1995; Murray et al., 2002; Yu et al., 2014; Han and Balakrishnan, 2002; Lendaris et al., 2000; Liu and Balakrishnan, 2000; Ferrari-Trecate et al., 2003] in automotive control, flight control, power control, among others. A review of recent techniques can be found in Kalyanakrishnan and Stone [2009], Busoniu et al. [2017], Recht [2019], Polydoros and Nalpantidis

[2017], and Kamalapurkar et al. [2018]. The Q-learning approach learns an action-dependent function using Temporal Difference (TD) to obtain the optimal policy. This is inherently a discrete approach. There are continuous extensions of this technique, such as [Millán et al., 2002; Gaskett et al., 1999; Ryu et al., 2019; Wei et al., 2018]. However, for an efficient implementation, the state and action spaces ought to be finite, which is highly restrictive for continuous problems.

Adaptive controllers [Åström and Wittenmark, 2013], as a well-known class of control techniques, may seem similar to RL in methodology, while there are substantial differences in the problem formulation and objectives. Adaptive techniques, as well as RL, learn to regulate unknown systems utilizing data collected in real time. In fact, an RL technique can be seen as an adaptive technique that converges to the optimal control [Lewis and Vrabie, 2009]. However, as opposed to RL and optimal controllers, adaptive controllers are not normally intended to be optimal, with respect to a user-specified cost function. Hence, we will not draw direct comparisons with such methods.

Value methods in RL normally require solving the well-known HJB. However, common techniques for solving such equations suffer from the curse of dimensionality. Hence, in approximate DP techniques, a parametric or nonparametric model is used to approximate the solution. In Lewis and Vrabie [2009], some related approaches are reviewed that fundamentally follow the actor-critic structure [Barto et al., 1983], such as VI and PI algorithms.

In such approaches, the Bellman error, which is obtained from the exploration of the state space, is used to improve the parameters estimated in a gradient-descent or least-squares loop that requires the Persistence of Excitation (PE) condition. Since the Bellman error obtained is only valid along the trajectories of the system, sufficient exploration in the state space is required to efficiently estimate the parameters. In Kamalapurkar et al. [2018], the authors have reviewed different strategies employed to increase the data efficiency in exploration. In Vamvoudakis et al. [2012] and Modares et al. [2014], a probing signal is added to the control to enhance the exploring properties of the policy. In another approach [Modares et al., 2014], the recorded data of explorations are used as a replay of experience to increase the data efficiency. Accordingly, the model obtained from identification is used to acquire more experience by doing simulation in an offline routine that decreases the need for visiting any point in the state space.

As an alternative method, considering a nonlinear control-affine system with a known input coupling function, the work Kamalapurkar et al. [2016b] used a parametric model to approximate the value function. Then, they employed a least-squares minimization technique to adjust the parameters according to the

Bellman error, which can be calculated at any arbitrary point of the state space using identified internal dynamics of the system and approximated state derivatives under a PE-like rank condition. In Kamalapurkar et al. [2016a], the authors proposed an improved technique, which approximates the value function only in a small neighborhood of the current state. It has been shown that the local approximation can be done more efficiently since a considerably less number of bases can be used.

In the work by Jiang and Jiang [2012], Jiang and Jiang [2014], and Jiang and Jiang [2017], the authors proposed PI-based algorithms that do not require any prior knowledge of the system model. A similar PE-like rank condition was used to ensure sufficient exploration for successful learning of the value functions and controllers. It is shown that these algorithms can achieve semiglobal stabilization and convergence to optimal values and controllers. One of the main limitations of PI-based algorithms is that an initial stabilizing controller has to be provided. While convergent PI algorithms have been recently proved for discrete-time systems in more general settings [Bertsekas, 2017], its extension to continuous-time systems involves substantial technical difficulties as pointed out by Bertsekas [2017].

I.2.5 Piecewise Learning

There exist different techniques to efficiently fit a piecewise model to data, see, e.g. Toriello and Vielma [2012], Breschi et al. [2016], Ferrari-Trecate et al. [2003], Amaldi et al. [2016], Rebennack and Krasko [2020], and Du et al. [2021]. In Ferrari-Trecate et al. [2003], a technique for the identification of discrete-time hybrid systems by the piecewise affine model is presented. The algorithm combines clustering, linear identification, and pattern recognition approaches to identify the affine subsystems together with the partitions for which they apply. In fact, the problem of globally fitting a piecewise affine model is considered to be computationally expensive to solve. In Lauer [2015], it is discussed that global optimality can be reached with a polynomial complexity in the number of data, while it is exponential with respect to the data dimension. In this regard, the work by Breschi et al. [2016] presents an efficient two-step technique: first, recursively clustering of the regressor vectors and estimation of the model parameters, and second, computation of a polyhedral partition. A review of some of the techniques can be found in Gambella et al. [2021] and Garulli et al. [2012].

The flexibility of Piecewise Affine (PWA) systems makes them suitable for different approaches in control. Hence, the control problem of piecewise systems is extensively studied in the literature; see, e.g. Marcucci and Tedrake [2019], Zou and Li [2007], Rodrigues and Boyd [2005], Baotic [2005], Strijbosch et al. [2020], Christophersen et al. [2005], and Rodrigues and How [2003]. Moreover, various applications can be found for PWA systems, including robotics [Andrikopoulos et al., 2013; Marcucci et al., 2017], automotive control [Borrelli et al., 2006; Sun et al., 2019], and power electronics [Geyer et al., 2008; Vlad et al., 2012]. In Zou and Li [2007], the robust MPC strategy is extended to PWA systems with polytopic uncertainty, where multiple PWA quadratic Lyapunov functions are employed for different vertices of the uncertainty polytope in different partitions. In another work by Marcucci and Tedrake [2019], hybrid MPC is formulated as a mixed-integer program to solve the optimal control problem for PWA systems. However, these techniques are only available in an open-loop form, which decreases their applicability for real-time control.

On the other hand, Deep Neural Network (DNN) offers an efficient technique for control in closed loop. However, one drawback of DNN-based control is the difficulty in stability analysis. This becomes even more challenging when PWA are considered. The work by Chen et al. [2020] suggested a sample-efficient technique for synthesizing a Lyapunov function for the PWA system controlled through a DNN in closed loop. In this approach, Analytic Center Cutting-Plane Method (ACCPM) [Goffin and Vial, 1993; Nesterov, 1995; Boyd and Vandenberghe, 2004] is first used for searching for a Lyapunov function. Then, this Lyapunov function candidate is verified on the closed-loop system using an MIQP. This approach relies on our knowledge of the exact model of the system. Hence, it cannot be directly implemented on an identified PWA system with uncertainty.

I.2.6 Tracking Control

For the learning-based tracking problem, several techniques can be found in the literature, in addition to some extensions presented for the techniques reviewed by Modares and Lewis [2014], Modares et al. [2015], Zhu et al. [2016], Yang et al. [2016], and Luo et al. [2016]. Modares and Lewis [2014] have developed an integral RL technique for linear systems based on PI algorithm, starting with an admissible initial controller. It has been shown that the optimal tracking controller converges to the Linear Quadratic Tracking (LQT) controller, with a partially unknown system. In Modares et al. [2015], an off-policy method is employed with three neural networks in an actor-critic-disturbance configuration to learn an H_∞-tracking controller for unknown nonlinear systems. Zhu et al. [2016] constructed an augmented system using the tracking error and the reference. Neural networks were employed, in an actor-critic structure, to approximate the value function and learn an optimal policy. In another neural network-based approach [Yang et al., 2016], a single network was used to approximate the value function, where classes of uncertain dynamics were assumed. In addition to the above approaches, there exist other similar ones in the literature. However, applications of RL in tracking control are not only limited to model-based techniques. For instance, Luo et al. [2016] suggests a critic-only Q-learning approach for tracking problems, which does not require solving the HJB equation.

I.2.7 Applications

As mentioned, there exist different applications of MBRL, as well as the optimal control, on real-world problems [Prokhorov, 2008; Ferrari-Trecate et al., 2003; Prokhorov et al., 1995; Murray et al., 2002; Yu et al., 2014; Han and Balakrishnan, 2002; Lendaris et al., 2000; Liu and Balakrishnan, 2000; Ferrari-Trecate et al., 2003]. Accordingly, we will later provide the detailed literature review for each of the applications including the quadrotor and the solar PV systems in Chapters 9 and 8, respectively.

Bibliography

Edoardo Amaldi, Stefano Coniglio, and Leonardo Taccari. Discrete optimization methods to fit piecewise affine models to data points. *Computers & Operations Research*, 75:214–230, 2016.

George Andrikopoulos, George Nikolakopoulos, Ioannis Arvanitakis, and Stamatis Manesis. Piecewise affine modeling and constrained optimal control for a pneumatic artificial muscle. *IEEE Transactions on Industrial Electronics*, 61(2):904–916, 2013.

Kai Arulkumaran, Marc Peter Deisenroth, Miles Brundage, and Anil Anthony Bharath. Deep reinforcement learning: A brief survey. *IEEE Signal Processing Magazine*, 34(6):26–38, 2017.

Karl J. Åström and Björn Wittenmark. *Adaptive Control*. Courier Corporation, 2013.

Christopher G. Atkeson and Juan Carlos Santamaria. A comparison of direct and model-based reinforcement learning. In *Proceedings of the International Conference on Robotics and Automation*, volume 4, pages 3557–3564. IEEE, 1997.

S. N. Balakrishnan, Jie Ding, and Frank L. Lewis. Issues on stability of ADP feedback controllers for dynamical systems. *IEEE Transactions on Systems, Man, and Cybernetics, Part B (Cybernetics)*, 38(4):913–917, 2008.

Mato Baotic. Optimal Control of Piecewise Affine Systems: A Multi-parametric Approach. PhD thesis, ETH Zurich, 2005.

Andrew G. Barto, Richard S. Sutton, and Charles W. Anderson. Neuronlike adaptive elements that can solve difficult learning control problems. *IEEE Transactions on Systems, Man, and Cybernetics*, SMC-13(5):834–846, 1983.

Richard E. Bellman and Stuart E. Dreyfus. *Applied Dynamic Programming*. Princeton University Press, 1962.

Dimitri P. Bertsekas. Value and policy iterations in optimal control and adaptive dynamic programming. *IEEE Transactions on Neural Networks and Learning Systems*, 28(3):500–509, 2017.

Francesco Borrelli, Alberto Bemporad, Michael Fodor, and Davor Hrovat. An MPC/hybrid system approach to traction control. *IEEE Transactions on Control Systems Technology*, 14(3):541–552, 2006.

Stephen Boyd and Lieven Vandenberghe. *Convex Optimization*. Cambridge University Press, 2004.

Valentina Breschi, Dario Piga, and Alberto Bemporad. Piecewise affine regression via recursive multiple least squares and multicategory discrimination. *Automatica*, 73:155–162, 2016.

Lucian Busoniu, Robert Babuska, Bart De Schutter, and Damien Ernst. *Reinforcement Learning and Dynamic Programming Using Function Approximators*. CRC Press, 2017.

Eduardo F. Camacho and Carlos Bordons Alba. *Model Predictive Control*. Springer, 2013.

Min-Shin Chen and Chung-yao Kao. Control of linear time-varying systems using forward Riccati equation. *Journal of Dynamic Systems, Measurement, and Control*, 119(3):536–540, 1997.

Shaoru Chen, Mahyar Fazlyab, Manfred Morari, George J. Pappas, and Victor M. Preciado. Learning Lyapunov functions for piecewise affine systems with neural network controllers. *arXiv preprint arXiv:2008.06546*, 2020.

Frank J. Christophersen, Mato Baotić, and Manfred Morari. Optimal control of piecewise affine systems: A dynamic programming approach. In *Control and Observer Design for Nonlinear Finite and Infinite Dimensional Systems*, pages 183–198. Springer, 2005.

Tayfun Çimen. State-dependent Riccati equation (SDRE) control: A survey. *IFAC Proceedings Volumes*, 41(2):3761–3775, 2008.

James R. Cloutier. State-dependent Riccati equation techniques: An overview. In *Proceedings of the American Control Conference*, volume 2, pages 932–936. IEEE, 1997.

Felipe Leno Da Silva and Anna Helena Reali Costa. A survey on transfer learning for multiagent reinforcement learning systems. *Journal of Artificial Intelligence Research*, 64:645–703, 2019.

Yingwei Du, Fangzhou Liu, Jianbin Qiu, and Martin Buss. Online identification of piecewise affine systems using integral concurrent learning. *IEEE Transactions on Circuits and Systems I: Regular Papers*, 68(10):4324–4336, 2021.

Yan Duan, Xi Chen, Rein Houthooft, John Schulman, and Pieter Abbeel. Benchmarking deep reinforcement learning for continuous control. In *International Conference on Machine Learning*, pages 1329–1338, 2016.

Gabriel Dulac-Arnold, Daniel Mankowitz, and Todd Hester. Challenges of real-world reinforcement learning. *arXiv preprint arXiv:1904.12901*, 2019.

Evrin B. Erdem and Andrew G. Alleyne. Design of a class of nonlinear controllers via state dependent Riccati equations. *IEEE Transactions on Control Systems Technology*, 12(1):133–137, 2004.

Giancarlo Ferrari-Trecate, Marco Muselli, Diego Liberati, and Manfred Morari. A clustering technique for the identification of piecewise affine systems. *Automatica*, 39(2):205–217, 2003.

Claudio Gambella, Bissan Ghaddar, and Joe Naoum-Sawaya. Optimization problems for machine learning: A survey. *European Journal of Operational Research*, 290(3):807–828, 2021.

Carlos E. Garcia, David M. Prett, and Manfred Morari. Model predictive control: Theory and practice–a survey. *Automatica*, 25(3):335–348, 1989.

Andrea Garulli, Simone Paoletti, and Antonio Vicino. A survey on switched and piecewise affine system identification. *IFAC Proceedings Volumes*, 45(16):344–355, 2012.

Chris Gaskett, David Wettergreen, and Alexander Zelinsky. Q-learning in continuous state and action spaces. In *Proceedings of the Australasian Joint Conference on Artificial Intelligence*, pages 417–428. Springer, 1999.

Tobias Geyer, Georgios Papafotiou, and Manfred Morari. Hybrid model predictive control of the step-down DC–DC converter. *IEEE Transactions on Control Systems Technology*, 16(6):1112–1124, 2008.

Jean-Louis Goffin and Jean-Philippe Vial. On the computation of weighted analytic centers and dual ellipsoids with the projective algorithm. *Mathematical Programming*, 60(1):81–92, 1993.

Lars Grüne and Jürgen Pannek. Nonlinear model predictive control. In *Nonlinear Model Predictive Control*, pages 45–69. Springer, 2017.

Dongchen Han and S. N. Balakrishnan. State-constrained agile missile control with adaptive-critic-based neural networks. *IEEE Transactions on Control Systems Technology*, 10(4):481–489, 2002.

Ammar Haydari and Yasin Yilmaz. Deep reinforcement learning for intelligent transportation systems: A survey. *IEEE Transactions on Intelligent Transportation Systems*, 2020.

Pablo Hernandez-Leal, Bilal Kartal, and Matthew E. Taylor. A survey and critique of multiagent deep reinforcement learning. *Autonomous Agents and Multi-Agent Systems*, 33(6):750–797, 2019.

Yu Jiang and Zhong-Ping Jiang. Computational adaptive optimal control for continuous-time linear systems with completely unknown dynamics. *Automatica*, 48(10):2699–2704, 2012.

Yu Jiang and Zhong-Ping Jiang. Robust adaptive dynamic programming and feedback stabilization of nonlinear systems. *IEEE Transactions on Neural Networks and Learning Systems*, 25(5):882–893, 2014.

Yu Jiang and Zhong-Ping Jiang. *Robust Adaptive Dynamic Programming*. John Wiley & Sons, 2017.

Leslie Pack Kaelbling, Michael L. Littman, and Andrew W. Moore. Reinforcement learning: A survey. *Journal of Artificial Intelligence Research*, 4:237–285, 1996.

Rudolf E. Kalman. Contributions to the theory of optimal control. *Boletin de la Sociedad Matematica Mexicana*, 5(2):102–119, 1960.

Shivaram Kalyanakrishnan and Peter Stone. An empirical analysis of value function-based and policy search reinforcement learning. In *Proceedings of the 8th International Conference on Autonomous Agents and Multiagent Systems - Volume 2*, pages 749–756, 2009.

Rushikesh Kamalapurkar, Joel A. Rosenfeld, and Warren E. Dixon. Efficient model-based reinforcement learning for approximate online optimal control. *Automatica*, 74:247–258, 2016a.

Rushikesh Kamalapurkar, Patrick Walters, and Warren E. Dixon. Model-based reinforcement learning for approximate optimal regulation. *Automatica*, 64(C):94–104, 2016b.

Rushikesh Kamalapurkar, Patrick Walters, Joel Rosenfeld, and Warren Dixon. *Reinforcement Learning for Optimal Feedback Control*. Springer, 2018.

B. Ravi Kiran, Ibrahim Sobh, Victor Talpaert, Patrick Mannion, Ahmad A. Al Sallab, Senthil Yogamani, and Patrick Pérez. Deep reinforcement learning for autonomous driving: A survey. *IEEE Transactions on Intelligent Transportation Systems*, 2021.

Jens Kober, J. Andrew Bagnell, and Jan Peters. Reinforcement learning in robotics: A survey. *The International Journal of Robotics Research*, 32(11):1238–1274, 2013.

Fabien Lauer. On the complexity of piecewise affine system identification. *Automatica*, 62:148–153, 2015.

George G. Lendaris, Larry Schultz, and Thaddeus Shannon. Adaptive critic design for intelligent steering and speed control of a 2-axle vehicle. In *Proceedings of the IEEE-INNS-ENNS International Joint Conference on Neural Networks. IJCNN 2000.*

xxxii Introduction

Neural Computing: New Challenges and Perspectives for the New Millennium,
volume 3, pages 73–78. IEEE, 2000.

Frank L. Lewis and Draguna Vrabie. Reinforcement learning and adaptive dynamic
programming for feedback control. *IEEE Circuits and Systems Magazine*,
9(3):32–50, 2009.

Frank L. Lewis, Draguna Vrabie, and Kyriakos G. Vamvoudakis. Reinforcement
learning and feedback control: Using natural decision methods to design optimal
adaptive controllers. *IEEE Control Systems Magazine*, 32(6):76–105, 2012.

Xin Liu and S. N. Balakrishnan. Convergence analysis of adaptive critic based optimal
control. In *Proceedings of the American Control Conference*, volume 3, pages
1929–1933. IEEE, 2000.

Biao Luo, Derong Liu, Tingwen Huang, and Ding Wang. Model-free optimal tracking
control via critic-only Q-learning. *IEEE Transactions on Neural Networks and
Learning Systems*, 27(10):2134–2144, 2016.

Tobia Marcucci and Russ Tedrake. Mixed-integer formulations for optimal control of
piecewise-affine systems. In *Proceedings of the ACM International Conference on
Hybrid Systems: Computation and Control*, pages 230–239, 2019.

Tobia Marcucci, Robin Deits, Marco Gabiccini, Antonio Bicchi, and Russ Tedrake.
Approximate hybrid model predictive control for multi-contact push recovery in
complex environments. In *Proceedings of the IEEE-RAS 17th International
Conference on Humanoid Robotics*, pages 31–38. IEEE, 2017.

David Q. Mayne and Hannah Michalska. Receding horizon control of nonlinear
systems. In *Proceedings of the IEEE Conference on Decision and Control*, pages
464–465. IEEE, 1988.

José Del R. Millán, Daniele Posenato, and Eric Dedieu. Continuous-action
Q-learning. *Machine Learning*, 49(2):247–265, 2002.

Hamidreza Modares and Frank L. Lewis. Linear quadratic tracking control of
partially-unknown continuous-time systems using reinforcement learning. *IEEE
Transactions on Automatic Control*, 59(11):3051–3056, 2014.

Hamidreza Modares, Frank L. Lewis, and Mohammad-Bagher Naghibi-Sistani.
Integral reinforcement learning and experience replay for adaptive optimal control
of partially-unknown constrained-input continuous-time systems. *Automatica*,
50(1):193–202, 2014.

Hamidreza Modares, Frank L. Lewis, and Zhong-Ping Jiang. H_∞ tracking control of
completely unknown continuous-time systems via off-policy reinforcement
learning. *IEEE Transactions on Neural Networks and Learning Systems*,
26(10):2550–2562, 2015.

Thomas M. Moerland, Joost Broekens, and Catholijn M. Jonker. Model-based
reinforcement learning: A survey. *arXiv preprint arXiv:2006.16712*, 2020.

Manfred Morari and Jay H. Lee. Model predictive control: Past, present and future. *Computers and Chemical Engineering*, 23(4–5):667–682, 1999.

John J. Murray, Chadwick J. Cox, George G. Lendaris, and Richard Saeks. Adaptive dynamic programming. *IEEE Transactions on Systems, Man, and Cybernetics, Part C (Applications and Reviews)*, 32(2):140–153, 2002.

Yu Nesterov. Complexity estimates of some cutting plane methods based on the analytic barrier. *Mathematical Programming*, 69(1):149–176, 1995.

Athanasios S. Polydoros and Lazaros Nalpantidis. Survey of model-based reinforcement learning: Applications on robotics. *Journal of Intelligent and Robotic Systems*, 86(2):153–173, 2017.

Lev Semenovich Pontryagin. *Mathematical Theory of Optimal Processes*. CRC Press, 1987.

Warren Buckler Powell. *Handbook of Learning and Approximate Dynamic Programming*, volume 2. John Wiley & Sons, 2004.

Anna Prach, Ozan Tekinalp, and Dennis S. Bernstein. Infinite-horizon linear-quadratic control by forward propagation of the differential Riccati equation. *IEEE Control Systems Magazine*, 35(2):78–93, 2015.

Danil Prokhorov. Neural networks in automotive applications. In *Computational Intelligence in Automotive Applications*, pages 101–123. Springer, 2008.

Danil V. Prokhorov, Roberto A. Santiago, and Donald C. Wunsch II. Adaptive critic designs: A case study for neurocontrol. *Neural Networks*, 8(9):1367–1372, 1995.

S. Joe Qin and Thomas A. Badgwell. A survey of industrial model predictive control technology. *Control Engineering Practice*, 11(7):733–764, 2003.

Steffen Rebennack and Vitaliy Krasko. Piecewise linear function fitting via mixed-integer linear programming. *INFORMS Journal on Computing*, 32(2):507–530, 2020.

Benjamin Recht. A tour of reinforcement learning: The view from continuous control. *Annual Review of Control, Robotics, and Autonomous Systems*, 2:253–279, 2019.

Luis Rodrigues and Stephen Boyd. Piecewise-affine state feedback for piecewise-affine slab systems using convex optimization. *Systems & Control Letters*, 54(9):835–853, 2005.

Luis Rodrigues and Jonathan P. How. Observer-based control of piecewise-affine systems. *International Journal of Control*, 76(5):459–477, 2003.

Moonkyung Ryu, Yinlam Chow, Ross Anderson, Christian Tjandraatmadja, and Craig Boutilier. CAQL: Continuous action Q-learning. In *Proceedings of the International Conference on Learning Representations*, 2019.

Nard Strijbosch, Isaac Spiegel, Kira Barton, and Tom Oomen. Monotonically convergent iterative learning control for piecewise affine systems. *IFAC-PapersOnLine*, 53(2):1474–1479, 2020.

Xiaoqiang Sun, Houzhong Zhang, Yingfeng Cai, Shaohua Wang, and Long Chen. Hybrid modeling and predictive control of intelligent vehicle longitudinal velocity considering nonlinear tire dynamics. *Nonlinear Dynamics*, 97(2):1051–1066, 2019.

Richard S. Sutton. Integrated architectures for learning, planning, and reacting based on approximating dynamic programming. In *Machine Learning Proceedings 1990*, pages 216–224. Elsevier, 1990.

Richard S. Sutton and Andrew G. Barto. *Reinforcement Learning: An Introduction*. MIT Press, 2018.

Alejandro Toriello and Juan Pablo Vielma. Fitting piecewise linear continuous functions. *European Journal of Operational Research*, 219(1):86–95, 2012.

Kyriakos G. Vamvoudakis, Frank L. Lewis, and Greg R. Hudas. Multi-agent differential graphical games: Online adaptive learning solution for synchronization with optimality. *Automatica*, 48(8):1598–1611, 2012.

Cristina Vlad, Pedro Rodriguez-Ayerbe, Emmanuel Godoy, and Pierre Lefranc. Explicit model predictive control of buck converter. In *Proceedings of the International Power Electronics and Motion Control Conference*, pages DS1e–4. IEEE, 2012.

Fei-Yue Wang, Huaguang Zhang, and Derong Liu. Adaptive dynamic programming: An introduction. *IEEE Computational Intelligence Magazine*, 4(2):39–47, 2009.

Ermo Wei, Drew Wicke, David Freelan, and Sean Luke. Multiagent soft Q-learning. In *2018 AAAI Spring Symposium Series*, 2018.

Avishai Weiss, Ilya Kolmanovsky, and Dennis S. Bernstein. Forward-integration Riccati-based output-feedback control of linear time-varying systems. In *Proceedings of the American Control Conference*, pages 6708–6714. IEEE, 2012.

Marco A. Wiering and Martijn Van Otterlo, editors. *Reinforcement Learning: State-of-the-Art*. Springer, 2012.

Xiong Yang, Derong Liu, Qinglai Wei, and Ding Wang. Guaranteed cost neural tracking control for a class of uncertain nonlinear systems using adaptive dynamic programming. *Neurocomputing*, 198:80–90, 2016.

Ting Yang, Liyuan Zhao, Wei Li, and Albert Y. Zomaya. Reinforcement learning in sustainable energy and electric systems: A survey. *Annual Reviews in Control*, 49:145–163, 2020.

Miao Yu, Chao Lu, and Yongjun Liu. Direct heuristic dynamic programming method for power system stability enhancement. In *2014 American Control Conference*, pages 747–752. IEEE, 2014.

Chao Yu, Jiming Liu, Shamim Nemati, and Guosheng Yin. Reinforcement learning in healthcare: A survey. *ACM Computing Surveys*, 55(1):1–36, 2021.

Zidong Zhang, Dongxia Zhang, and Robert C. Qiu. Deep reinforcement learning for power system applications: An overview. *CSEE Journal of Power and Energy Systems*, 6(1):213–225, 2019.

Kaiqing Zhang, Zhuoran Yang, and Tamer Başar. Multi-agent reinforcement learning: A selective overview of theories and algorithms. In *Handbook of Reinforcement Learning and Control*, pages 321–384, 2021.

Yuanheng Zhu, Dongbin Zhao, and Xiangjun Li. Using reinforcement learning techniques to solve continuous-time non-linear optimal tracking problem without system dynamics. *IET Control Theory and Applications*, 10(12):1339–1347, 2016.

Yuanyuan Zou and Shaoyuan Li. Robust model predictive control for piecewise affine systems. *Circuits, Systems & Signal Processing*, 26(3):393–406, 2007.

1

Nonlinear Systems Analysis

In this chapter, we give a basic mathematical introduction to analysis of nonlinear dynamical systems, with a focus on Lyapunov stability analysis.

Stability analysis is known as one of the main challenges in the study of dynamical systems. The responses of the system against perturbations can demonstrate the quality of stability in the system. Considering the behavior of the system around an equilibrium point, different scenarios are expected for the trajectories of the system under perturbations. Accordingly, different notions of stability arise. In this regard, Lyapunov analysis provides an efficient framework for analyzing the stability of nonlinear dynamical systems. Moreover, it can aid in designing feedback controllers as well as analyzing the closed-loop system. In this chapter, we briefly review some basic concepts of stability and discuss several well-known Lyapunov stability theorems in a self-contained manner. For detailed discussions, we refer the readers to Haddad and Chellaboina [2011] and Khalil [2002].

1.1 Notation

We introduce a minimal set of mathematical notation that will be used in this book.

- $\mathbb{R}$: The set of real numbers.
- $\mathbb{R}_{\geq 0}$: The set of nonnegative real numbers.
- $\mathbb{Z}$: The set of all integers.
- $\mathbb{Z}_{\geq 0}$: The set of nonnegative integers.
- $|x|$: The absolute value of a real number.
- $\mathbb{R}^n$: The n-dimensional Euclidean space.
- $\|x\|$: The 2-norm (or Euclidean norm) of a vector $x \in \mathbb{R}^n$, defined by

$$\|x\| = \sqrt{\sum_{i=1}^{n} |x_i|^2}.$$

Model-Based Reinforcement Learning: From Data to Continuous Actions with a Python-based Toolbox, First Edition. Milad Farsi and Jun Liu.

- $A\backslash B$: The set difference between the set A and the set B, defined by

$$A\backslash B = \{x \mid x \in A, \; x \notin B\}.$$

- $\|x\|_A$: The distance from a point $x \in \mathbb{R}^n$ to a set $A \subseteq \mathbb{R}^n$, i.e.

$$\|x\|_A = \inf_{y \in A} \|x - y\|.$$

- $B_r(A)$: The set consisting of all points with distance less than or equal to r from the set $A \subseteq \mathbb{R}^n$, i.e.

$$B_r(A) = \{x : \; \|x\|_A \leq r\}.$$

If $A = \{x\}$, where $x \in \mathbb{R}^n$, $B_r(A)$ reduces to $B_r(x)$. If $x = 0$, we simply write it as B_r.

- $\frac{\partial V}{\partial x}$: The gradient for a function $V : \; \mathbb{R}^n \to \mathbb{R}$ defined by $\frac{\partial V}{\partial x} = \left(\frac{\partial V}{\partial x_1}, \ldots, \frac{\partial V}{\partial x_n} \right)$.

- $\frac{\partial \Phi}{\partial x}$: The Jacobian matrix for a function $\Phi : \; \mathbb{R}^n \to \mathbb{R}^p$ defined by

$$\frac{\partial \Phi}{\partial x} = \begin{bmatrix} \frac{\partial \Phi_1}{\partial x} \\ \vdots \\ \frac{\partial \Phi_p}{\partial x} \end{bmatrix} = \begin{bmatrix} \frac{\partial \Phi_1}{\partial x_1} & \cdots & \frac{\partial \Phi_1}{\partial x_n} \\ \vdots & \cdots & \vdots \\ \frac{\partial \Phi_p}{\partial x_1} & \cdots & \frac{\partial \Phi_p}{\partial x_n} \end{bmatrix}.$$

1.2 Nonlinear Dynamical Systems

We consider nonlinear dynamical systems of the form

$$\dot{x}(t) = f(t, x(t), w(t)), \quad x(t_0) = x_0, \tag{1.1}$$

where $x(t) \in \mathbb{R}^n$ is the system state, $w(t) \in W \subseteq \mathbb{R}^p$ is a disturbance signal, $t_0 \in \mathbb{R}_{\geq 0}$ is the initial time, and x_0 is the initial state. Let $D \subseteq \mathbb{R}^n$ be an open set and $J \subseteq \mathbb{R}$ be an open interval. We assume that $f : \; J \times D \times W \to \mathbb{R}^n$ satisfies the basic regularity conditions such that, for any $(t_0, x_0) \in J \times D$ and any "well-behaved" input signal $w : \; J \to W$, there exists some interval $J_0 \subseteq J$ containing t_0 and a unique local solution $x : \; J_0 \to \mathbb{R}^n$ such that (1.1) is satisfies for all $t \in J_0$.

1.2.1 Remarks on Existence, Uniqueness, and Continuation of Solutions

Remark 1.1 (*Basic regularity assumptions*) The dependence of f in t sometimes comes from obtaining f from another function $F(x, u, w)$, where u is a control input and w is a disturbance input. For input signals (either control or disturbance signals), they are usually assumed to be piecewise continuous. In this setting, we assume that $f(t, x, w)$ is piecewise continuous in t and continuous in x and w to ensure existence of a local solution, for a given piecewise continuous input $w(\cdot)$. This is known as *Peano's existence theorem*. If we further assume that f is locally Lipschitz continuous in x, i.e. for each bounded set $K \subseteq J \times D \times W$, there exists a constant L, such that

$$\|f(t, x, w) - f(t, y, w)\| \le L \|x - y\|, \quad \forall (t, x, w), (t, y, w) \in K,$$

then every solution to (1.1) is also unique. Existence and uniqueness under the Lipschitz continuity assumption is often known as *Picard's existence theorem*.

In a more general setting, we are sometimes required to consider input signals that are only measurable with regard to (w.r.t.) t. If we allow the input signals to measurable functions, then we need to relax the condition on f to be measurable w.r.t. t and allow $w(\cdot)$ to be locally essentially bounded, i.e. a measurable function that is "almost" equal to a function that is bounded on a neighborhood of every point. Here, "almost" means almost everywhere, i.e. except on a set of zero measure. *Carathéodory's existence theorem* deals with existence and uniqueness of solutions under this setting.

Remark 1.2 **(*Continuation of solutions*)** A local solution to (1.1) can be extended to its *maximum interval of existence* $J^* \subseteq J$. This interval can be shown to be open relative to J. Consider the special case of $J = \mathbb{R}$ and $D = \mathbb{R}^n$. Then the maximum interval of existence is $\mathbb{R}$, unless the solution *blows up* in finite time, i.e. the solution becomes unbounded as time approaches the boundary of the interval of existence. Solutions that are defined on $\mathbb{R}$ are called *global solutions*. If a solution is defined on $[t_0, \infty)$, we say it is *forward complete*. Similarly, if it is defined on $(-\infty, t_0]$, we say it is *backward complete*. Hence, a usual way to show global existence and completeness of solutions is to show that solutions remain bounded on bounded intervals.

We refer the readers to the Appendix of Sontag [2013] for a precise mathematical treatment of the basic theory of nonlinear systems of Ordinary Differential Equations (ODEs) with inputs (see also [Haddad and Chellaboina, 2011; Khalil, 2002, 2015]).

1.3 Lyapunov Analysis of Stability

Stability is a central notion in systems and control. We define stability for solutions of (1.1] with respect to a compact invariant set as follows.

Definition 1.1 A set $A \subseteq \mathbb{R}^n$ is said to be **(forward) invariant set** for system (1.1), if it is nonempty and, for any $x_0 \in A$, all solutions of (1.1) starting from $x(t_0) = x_0$ stay in A for all $t \ge t_0$.

Definition 1.2 Let $A \subseteq \mathbb{R}^n$ be a compact invariant set for system (1.1). We say that A is **Uniformly Asymptotically Stable (UAS)** for system (1.1), if the following two conditions hold:

1. (Uniform stability) For every $\varepsilon > 0$, there exists $\delta = \delta(\varepsilon) > 0$ such that, if $\|x_0\|_A \leq \delta$, then $\|x(t)\|_A \leq \varepsilon$ for all $t \geq t_0$, where $x(t)$ is any solution of (1.1) starting from $x(t_0) = x_0$.
2. (Uniform attraction) There exists some $\rho > 0$ such that, for any $\eta > 0$, there exists some $T = T(\rho, \eta) \geq 0$ such that $\|x(t)\|_A \leq \eta$, whenever $\|x_0\|_A \leq \rho$ and $t \geq t_0 + T$, where $x(t)$ is any solution of (1.1) starting from $x(t_0) = x_0$.

We say that A is **Globally Uniformly Asymptotically Stable (GUAS)** for system (1.1), if the above conditions hold for δ chosen such that $\lim_{\varepsilon \to \infty} \delta(\varepsilon) = \infty$ and any $\rho > 0$.

We also introduce a special case of asymptotic stability as follows.

Definition 1.3 Let $A \subseteq \mathbb{R}^n$ be a compact invariant set for system (1.1). We say that A is **Uniformly Exponentially Stable (UES)** for system (1.1), if then there exist positive constants ρ, k, and c such that

$$\|x(t)\|_A \leq k\|x_0\|_A e^{-c(t-t_0)}, \quad \forall t \geq t_0, \forall \|x_0\|_A \leq \rho,$$

holds for all solutions of (1.1). It is said to be **Globally Uniformly Exponentially Stable (GUES)** if the above holds for all $x_0 \in \mathbb{R}^n$.

Remark 1.3 Note that, when $A = \{0\}$ and $x = 0$ is an equilibrium point for system (1.1), i.e. $f(t, 0, w) = 0$ for all $t \geq 0$ and $w \in W$, the above notions of stability reduce to the corresponding notions of stability for an equilibrium.

We present a standard Lyapunov theorem on stability analysis of system (1.1) with respect to a compact invariant set.

Definition 1.4 Let $D \subseteq \mathbb{R}^n$ be an open set and A be a compact set contained in D. We say that a function $V : D \to \mathbb{R}$ is **positive definite with respect to** A if $V(x) = 0$ for all $x \in A$ and $V(x) > 0$ for all $x \in D \backslash A$. Similarly, we say that $V : D \to \mathbb{R}$ is negative definite with respect to A, if $-V$ is positive definite with respect to A.

Definition 1.5 We say that a function $W : \mathbb{R}^n \to \mathbb{R}$ is **radially unbounded** if $W(x) \to \infty$ as $\|x\| \to \infty$.

Theorem 1.1 *(Lyapunov theorem for stability)* *Let A be a compact invariant set of system (1.1). Let $D \subseteq \mathbb{R}^n$ be an open set containing A and $V : [0, \infty) \times D \to \mathbb{R}$*

be a continuously differentiable function. Suppose that there exist continuous functions W_i ($i = 1,2,3$) that are defined on D and positive definite with respect to A such that

$$W_1(x) \leq V(t,x) \leq W_2(x), \quad \forall x \in D, \quad \forall t \geq 0, \tag{1.2}$$

and

$$\frac{dV}{dt} + \frac{dV}{dx} f(t,x,w) \leq -W_3(x), \quad \forall x \in D\backslash A, \quad \forall t \geq 0, \quad \forall w \in W. \tag{1.3}$$

Then A is UAS for system (1.1). If the above conditions hold for $D = \mathbb{R}^n$ and W_1 is radially unbounded, then A is GUAS for system (1.1).

Proof: (**Uniform stability**) Fix an arbitrary $\varepsilon > 0$. Without loss of generality, assume that ε is sufficiently small such that $B_\varepsilon(A) \subseteq D$. Let $c < \min_{\{x: \|x\|_A = \varepsilon\}} W_1(x)$. Then the set

$$W_1^c := \left\{ x \in B_\varepsilon(A) : \ W_1(x) \leq c \right\}$$

is contained in the interior of $B_\varepsilon(A)$. Define

$$W_2^c := \left\{ x \in B_\varepsilon(A) : \ W_2(x) \leq c \right\}.$$

Then $W_2^c \subseteq W_1^c$. Pick $\delta \in (0, \varepsilon)$ such that $B_\delta(A) \subseteq W_2^c$. This is always possible because $W_2(x)$ is continuous on D and $W_2(x) = 0$ for all $x \in A$. Hence, for sufficiently small $\delta > 0$, $W_2(x) \leq c$ for all $x \in B_\delta(A)$, which implies $B_\delta(A) \subseteq W_2^c$. We claim that solutions of (1.1) starting from any initial state $x_0 \in B_\delta(A)$ and any initial time $t_0 \geq 0$ will remain in $W_1^c \subseteq B_\varepsilon(A)$ for all $t \geq t_0$. This would imply uniform stability.

Pick any $x_0 \in B_\delta(A) \subseteq W_2^c$. Let $x(t)$ be any solution of (1.1) satisfying $x(t_0) = x_0$. Then

$$V(t_0, x(t_0)) = V(t_0, x_0) \leq W_2(x_0) \leq c.$$

We have

$$\frac{dV(t,x(t))}{dt} = \frac{\partial V}{\partial t} + \frac{dV}{dx} f(t,x(t),w(t)) \leq -W_3(x(t)) \leq 0 \tag{1.4}$$

for all $t \geq t_0$, provided that $x(t)$ remains in D. To escape $B_\varepsilon(A)$, the solution needs to cross the boundary of $B_\varepsilon(A)$, on which $W_1(x)) > c$. This implies that $V(t,x(t)) \geq W_1(x(t)) > c$ for some t. This is impossible, since (1.4) implies that $V(t,x(t))$ is non-increasing for $x(t) \in D$.

(**Uniform attraction**) From the proof of uniform stability, we have shown that, for every $\varepsilon > 0$ such that $B_\varepsilon(A) \subseteq D$, there exists some $\delta > 0$ such that solutions of (1.1) starting from $B_\delta(A)$ will stay in $B_\varepsilon(A)$ for all $t \geq t_0$. We fix some choice of ε and δ for the following argument. Let $\rho = \delta$. Fix any $\eta \in (0, \varepsilon)$ (without loss of generality). We show that there exists $T = T(\eta)$ such that any solution $x(t)$ of (1.1) starting in $B_\delta(A)$ will reach and stay in $B_\eta(A)$ for all $t \geq t_0 + T$. By the argument of uniform stability again, there exists some $\delta' = \delta'(\eta) > 0$ such that solutions of (1.1) starting in $B_{\delta'}(A)$ will stay in $B_\eta(A)$ for all future time. Hence, we only need to show that solutions starting in $B_\delta(A)$ will reach $B_{\delta'}(A)$ within some finite time T.

Let $\lambda = \min_{\{x: \ \delta' \leq \|x\|_A \leq \varepsilon\}} W_3(x) > 0$. Let $c = \max_{x \in B_\varepsilon(A)} W_2(x)$. Choose $T > \frac{c}{\lambda}$. Let $x(t)$ be any solution of (1.1) starting from $x(t_0) = x_0 \in B_\rho(A)$. Without loss of generality, assume $x_0 \notin B_\delta(A)$. Otherwise, $x(t) \in B_\eta(A)$ for all $t \geq t_0$.

Suppose that $x(t)$ never reaches $B_{\delta'}(A)$ on $[t_0, t_0 + T]$. That is, $x(t) \in \{x : \delta' \leq \|x\|_A \leq \varepsilon\}$ for all $t \in [t_0, t_0 + T]$. Then we have

$$W_1(x(t_0 + T))$$
$$\leq V(t_0 + T, x(t_0 + T))$$
$$= V(t_0, x_0) + \int_{t_0}^{t_0+T} \frac{dV(s, x(s))}{ds} ds$$
$$= V(t_0, x_0) + \int_{t_0}^{t_0+T} \left[\frac{\partial V}{\partial t}(s, x(s)) + \frac{dV}{dx}(s, x(s)) f(s, x(s), w(s)) \right] ds$$
$$\leq V(t_0, x_0) - \int_{t_0}^{t_0+T} W_3(x(s)) ds$$
$$\leq W_2(x_0) - \lambda T \leq c - \lambda T < 0, \tag{1.5}$$

which is a contradiction because $W_1(x)$ cannot be negative. Hence, $x(t)$ must have reached $B_{\delta'}(A)$ for some $t \in [t_0, t_0 + T]$. It follows that $x(t) \in B_\eta(A)$ for all $t \geq t_0 + T$.

(**Global stability**) When $D = \mathbb{R}^n$ and $W_1(x)$ is radially unbounded, we can pick $\delta = \delta(\varepsilon)$ for uniform stability such that $\lim_{\varepsilon \to \infty} \delta(\varepsilon) = \infty$. This is because as $c \to \infty$, as $\varepsilon \to \infty$, and W_2^c can contain any given $B_\delta(A)$ for c sufficiently large. ∎

The next theorem deals with exponential stability.

Theorem 1.2 (*Lyapunov theorem for exponential stability*) *Let A be a compact invariant set of system (1.1). Let $D \subseteq \mathbb{R}^n$ be an open set containing A and $V : [0, \infty) \times D \to \mathbb{R}$ be a continuously differentiable function. Suppose that there exist constants c_i ($i = 1, 2, 3$) and p such that*

$$c_1 \|x\|_A^p \leq V(t, x) \leq c_2 \|x\|_A^p, \quad \forall x \in D, \quad \forall t \geq 0, \tag{1.6}$$

and

$$\frac{dV}{dt} + \frac{dV}{dx} f(t, x, w) \leq -c_3 \|x\|_A^p, \quad \forall x \in D \backslash A, \quad \forall t \geq 0, \quad \forall w \in W. \tag{1.7}$$

Then A is UES for system (1.1). If the above conditions hold for $D = \mathbb{R}^n$, then A is GUES for system (1.1).

Proof: By Theorem 1.1, for any $\varepsilon > 0$ such that $B_\varepsilon(A) \subseteq D$, there exists some $\delta \in (0, \varepsilon)$ such that solutions starting from $B_\delta(A)$ remain in $B_\varepsilon(A)$ for all $t \geq t_0$. Let $x(t)$ be a solution. By the condition (1.7), we have

$$\frac{dV(t, x(t))}{dt} = \frac{\partial V}{\partial t}(t, x(t)) + \frac{dV}{dx}(t, x(t))f(t, x(t), w(t))$$

$$\leq -c_3 \|x(t)\|_A^p$$

$$\leq -\frac{c_3}{c_2} V(t, x(t)),$$

which implies

$$V(t, x(t)) \leq V(t_0, x_0)e^{-\frac{c_3}{c_2}(t-t_0)}.$$

By condition (1.6), the above inequality implies

$$\|x(t)\|_A \leq \left[\frac{V(t, x(t))}{c_1} \right]^{\frac{1}{p}}$$

$$\leq \left[\frac{V(t_0, x_0)e^{-\frac{c_3}{c_2}(t-t_0)}}{c_1} \right]^{\frac{1}{p}}$$

$$\leq \left[\frac{c_2 \|x(t)\|_A^p \, e^{-\frac{c_3}{c_2}(t-t_0)}}{c_1} \right]^{\frac{1}{p}} = \left(\frac{c_2}{c_1} \right)^{\frac{1}{p}} \|x_0\|_A e^{-\frac{c_3}{c_2 p}(t-t_0)}, \quad t \geq t_0,$$

provided that $x_0 \in B_\delta(A)$. This shows that A is UES for system (1.1). If $D = \mathbb{R}^n$, then the above inequality holds for all $x_0 \in \mathbb{R}^n$, which implies that A is GUES. ∎

1.4 Stability Analysis of Discrete Time Dynamical Systems

In many situations, we also need to consider discrete time dynamical systems. For example, in control applications, a discrete time dynamical system arises when we compute the control inputs at discrete time instants. Discrete time dynamical systems can also be obtained as time discretization of continuous time dynamical systems. Such models are more amenable to sequential decision-making.

Consider a discrete time dynamical system of the form

$$x(t + 1) = f(t, x(t), w(t)), \quad x(t_0) = x_0, \tag{1.8}$$

where $x(t) \in \mathbb{R}^n$ is the system state, $w(t) \in W \subseteq \mathbb{R}^p$ is a disturbance signal, $t_0 \in \mathbb{Z}_{\geq 0}$ is the initial time, and x_0 is the initial state. In other words, compared with a continuous time dynamical system, the time t now lies in $\mathbb{Z}$ and the dynamics are described by the difference Eq. (1.8), instead of a differential equation. Given an initial condition $x(t_0) = x_0$, a sequence $\{w(t)\}_{t=t_0}^{\infty}$, a **solution** to (1.8) is a sequence $\{x(t)\}_{t=t_0}^{\infty}$ that satisfies (1.8). For solutions to be defined for all $t \geq t_0$, we assume that $f : \mathbb{R}_{\geq 0} \times D \times W \to D$.

We present similar results on Lyapunov analysis of discrete time dynamical systems of the form (1.8). The definitions are almost identical with that for continuous time systems.

Definition 1.6 A set $A \subseteq \mathbb{R}^n$ is said to be **(forward) invariant set** for system (1.8), if it is nonempty and, for any $x_0 \in A$, all solutions of (1.8) starting from $x(t_0) = x_0$ stay in A for all $t \geq t_0$.

Definition 1.7 Let $A \subseteq \mathbb{R}^n$ be a compact invariant set for system (1.8). We say that A is UAS for system (1.8), if the following two conditions hold:

1. (Uniform stability) For every $\varepsilon > 0$, there exists $\delta = \delta(\varepsilon) > 0$ such that, if $\|x_0\|_A \leq \delta$, then $\|x(t)\|_A \leq \varepsilon$ for all $t \geq t_0$, where $x(t)$ is any solution of (1.8) starting from $x(t_0) = x_0$.
2. (Uniform attraction) There exists some $\rho > 0$ such that, for any $\eta > 0$, there exists some $T = T(\rho, \eta) \geq 0$ such that $\|x(t)\|_A \leq \eta$, whenever $\|x_0\|_A \leq \rho$ and $t \geq t_0 + T$, where $x(t)$ is any solution of (1.8) starting from $x(t_0) = x_0$.

We say that A is GUAS for system (1.8), if the above conditions hold for δ chosen such that $\lim_{\varepsilon \to \infty} \delta(\varepsilon) = \infty$ and any $\rho > 0$.

Definition 1.8 Let $A \subseteq \mathbb{R}^n$ be a compact invariant set for system (1.8). We say that A is UES for system (1.8), if there exist positive constants ρ, k, and $\lambda \in (0,1)$ such that

$$\|x(t)\|_A \leq k\|x_0\|_A \lambda^{t-t_0}, \quad \forall t \geq t_0, \forall \|x_0\|_A \leq \rho,$$

holds for all solutions of (1.8). It is said to be GUES if the above holds for all $x_0 \in \mathbb{R}^n$.

Lyapunov functions can also be used to analyze the stability of discrete time systems.

Theorem 1.3 *(Lyapunov theorem for stability)* *Let A be a compact invariant set of system (1.8). Let $D \subseteq \mathbb{R}^n$ be an open set containing A and $V : [0, \infty) \times D \to \mathbb{R}$ be a continuous function. Suppose that there exist continuous functions W_i $(i = 1,2,3)$ that are defined on D and positive definite with respect to A such that*

$$W_1(x) \leq V(t,x) \leq W_2(x), \quad \forall x \in D, \quad \forall t \geq 0, \tag{1.9}$$

and

$$V(t+1, f(t,x,w)) - V(t,x) \leq -W_3(x), \quad \forall x \in D, \quad \forall t \geq 0, \quad \forall w \in W.$$
(1.10)

Then A is UAS for system (1.8). If the above conditions hold for $D = \mathbb{R}^n$, and W_1 is radially unbounded, then A is GUAS for system (1.8).

Proof: The proof follows almost verbatim the proof of Theorem 1.1, with (1.4) replaced by

$$V(t+1, x(t+1)) - V(t, x(t)) \leq -W_3(x(t)) \leq 0,$$
(1.11)

and (1.5) replaced by

$$
\begin{aligned}
W_1(x(t_0 + T)) &\leq V(t_0 + T, x(t_0 + T)) \\
&= V(t_0, x_0) + \sum_{s=t_0}^{t_0+T-1} [V(s+1, x(s+1)) - V(s, x(s))] \\
&\leq V(t_0, x_0) - \sum_{s=t_0}^{t_0+T-1} W_3(x(s)) ds \\
&\leq W_2(x_0) - \lambda T \leq c - \lambda T < 0.
\end{aligned}
$$
(1.12)

The conclusion follows. ∎

The next theorem concerns exponential stability of discrete time dynamical systems.

Theorem 1.4 **(Lyapunov theorem for exponential stability)** *Let A be a compact invariant set of system (1.8). Let $D \subseteq \mathbb{R}^n$ be an open set containing A and $V : [0, \infty) \times D \to \mathbb{R}$ be a continuous function. Suppose that there exist positive constants c_i ($i = 1, 2, 3$) and p such that $c_3 \in (0,1)$,*

$$c_1 \|x\|_A^p \leq V(t,x) \leq c_2 \|x\|_A^p, \quad \forall x \in D, \quad \forall t \geq 0,$$
(1.13)

and

$$V(t+1, f(t,x,w)) \leq c_3 V(t,x), \quad \forall x \in D, \quad \forall t \geq 0, \quad \forall w \in W.$$
(1.14)

Then A is UES for system (1.8). If the above conditions hold for $D = \mathbb{R}^n$ and W_1 is radially unbounded, then A is GUES for system (1.8).

Proof: Similar to the proof of Theorem 1.2, we can obtain

$$V(t, x(t)) \leq V(t_0, x_0) c_3^{t-t_0}.$$

which implies

$$\|x(t)\|_A \leq \left[\frac{V(t, x(t))}{c_1} \right]^{\frac{1}{p}}$$

$$\leq \left[\frac{V(t_0, x_0) c_3^{t-t_0}}{c_1} \right]^{\frac{1}{p}}$$

$$\leq \left[\frac{c_2 \|x(t)\|_A^p c_3^{t-t_0}}{c_1} \right]^{\frac{1}{p}} = \left(\frac{c_2}{c_1} \right)^{\frac{1}{p}} \|x_0\|_A \left(c_3^{1/p} \right)^{t-t_0}, \quad t \geq t_0.$$

The conclusion follows. ∎

1.5 Summary

The material of this chapter is standard Lyapunov stability analysis for nonlinear systems. For more comprehensive treatment and further reading, readers are referred to Khalil [2002] and Haddad and Chellaboina [2011]. Here we present the stability analysis with respect to a compact set, which is often the case when the system is perturbed by a nonvanishing uncertainty. We also discuss stability analysis of discrete time systems, which will be used in a later chapter for analyzing a time discretization of a learned dynamic model.

Bibliography

Wassim M. Haddad and VijaySekhar Chellaboina. *Nonlinear Dynamical Systems and Control: A Lyapunov-Based Approach*. Princeton University Press, 2011.

Hassan K. Khalil. *Nonlinear Systems*. Patience Hall, 2002.

Hassan K. Khalil. *Nonlinear Control*. Pearson, 2015.

Eduardo D. Sontag. *Mathematical Control Theory: Deterministic Finite Dimensional Systems*. Springer, 2013.

2

Optimal Control

Control methods in applications generally involve some trial and error processes through which design parameters are chosen to satisfy a desired performance measure of the system. Desirable performance is usually determined in terms of systems' responses such as peak overshoot, settling time, and rise time. Furthermore, to reach the desirable response of the system, the control effort usually needs to be observed and restricted in some domain. Such design considerations cannot be generally accomplished for complicated systems by classical control methods. Hence, the optimal control framework is well known as a direct approach to the synthesis of such systems. In this regard, the goal of optimal control theory is to obtain the control input that is required to satisfy a performance measure and physical constraints.

2.1 Problem Formulation

First, to define an optimal control problem, we need a model of the control system. Hence, consider the general nonlinear system given by

$$\dot{x} = f(t, x, u), \qquad x(t_0) = x_0, \tag{2.1}$$

where $x \in \mathbb{R}^n$ is the state, u is the control input taking values in $U \subseteq \mathbb{R}^m$, x_0 is the initial state at the initial time t_0. Existence and uniqueness of solutions follow the same discussion as in Remark 1.1 for system (1.1), with u replacing the input w there.

Then, a cost functional is required to assign a cost to any trajectory of the system. Behaviors of the control system are determined by the control signal. Hence, the cost functional depends on the control signal as well. This also allows

Model-Based Reinforcement Learning: From Data to Continuous Actions with a Python-based Toolbox, First Edition. Milad Farsi and Jun Liu.

us to directly penalize control signals that are not desirable (e.g. those exceeding a certain magnitude). A general form of cost functional can be written as follows:

$$J(t_0, x_0, t_1, u) = \int_{t_0}^{t_1} L(t, x(t), u(t))dt + K(t_1, x_1), \tag{2.2}$$

where L and K are the running cost and the terminal cost, respectively, and $x_1 = x(t_1)$ is the terminal state and t_1 is the terminal time. Given the cost functional, we can then define the optimal control problem as finding a control u such that the system takes a trajectory that minimizes the cost functional. This problem is known as the *Bolza problem* in optimal control, which formally includes the *Lagrange problem* as a special case where there is no terminal cost [Liberzon, 2011]. Moreover, this cost functional can represent different problems in control by introducing a target set as the final condition, i.e. $(t_1, x_1) \in S$ for some *target set* $S \subseteq [t_0, \infty) \times \mathbb{R}^n$. For instance, a free-time fixed-endpoint optimal problem is given by enforcing the target set $[t_0, \infty) \times \{x_1\}$.

2.2 Dynamic Programming

Consider the cost functional (2.2) with the fixed final time t_1, where the endpoint x_1 is free. The basic idea of *dynamic programming* (DP) is to embed the problem we would like to solve – in this case, the optimal control problem described above – in a larger class of problems and solve all problems at once. To this end, instead of $J(t_0, x_0, u)$,[1] DP suggests minimizing

$$J(t, x, u) = \int_{t}^{t_1} L(s, x(s), u(s))ds + K(t_1, x(t_1)),$$

where $t \in [t_0, t_1)$ and $x \in \mathbb{R}^n$. Clearly, this recovers the original cost by considering $(t, x) = (t_0, x_0)$.

Let $U^{[t, t_1]}$ denote the set of all *admissible* control inputs defined on $[t, t_1]$ and taking values in U. Define the *value function* by

$$V(t, x) = \inf_{u \in U^{[t, t_1]}} J(t, x, u). \tag{2.3}$$

The intuitive meaning of the value function is then the optimal *cost-to-go* from (t, x). We use an infimum instead of a minimum because we do not know *a priori* whether an optimal control that renders the cost-to-go optimal exists.

2.2.1 Principle of Optimality

DP aims to break a larger problem into smaller subproblems. This is enabled by the principle of optimality due to Bellman [1957].

---

1 Sine t_1 is fixed, we write $J(t_0, x_0, t_1, u)$ as $J(t_0, x_0, u)$.

Proposition 2.1 *(Bellman's principle of optimality)* For every $(t,x) \in [t_0, t_1) \times \mathbb{R}^n$ and every $\Delta > 0$ such that $t + \Delta < t_1$, the value function V defined in (2.3) satisfies

$$V(t,x) = \inf_{u \in U^{[t,t+\Delta]}} \left\{ \int_t^{t+\Delta} L(s,x(s),u(s))ds + V(t+\Delta, x(t+\Delta)) \right\}, \qquad (2.4)$$

where $x(\cdot)$ solves (2.1) on $[t, t + \Delta]$ with the control $u \in U^{[t,t+\Delta]}$ and $x(t) = x$.

The intuition behind the principle of optimality is that, for a control to be optimal on a larger time interval, it should also be optimal on every small interval, with the same running cost and optimal cost-to-go from the final time/state of the small time interval as the terminal cost.

Proof: Denote the right-hand side of (2.4) by V_Δ. We first show that $V(t,x) \geq V_\Delta$. By the definition of V, for any $\varepsilon > 0$, there exists $u_\varepsilon \in U^{[t,t_1]}$ such that

$$\int_t^{t_1} L(s,x(s),u_\varepsilon(s))ds + K(t_1,x(t_1)) \leq V(t,x) + \varepsilon,$$

where $x(\cdot)$ is a solution under control u_ε and with $x(t) = x$. It follows that

$$\begin{aligned}
V(t,x) + \varepsilon &\geq \int_t^{t_1} L(s,x(s),u_\varepsilon(s))ds + K(t_1,x(t_1)) \\
&= \int_t^{t+\Delta} L(s,x(s),u_\varepsilon(s))ds \\
&\quad + \int_{t+\Delta}^{t_1} L(s,x(s),u_\varepsilon(s))ds + K(t_1,x(t_1)) \\
&\geq \int_t^{t+\Delta} L(s,x(s),u_\varepsilon(s))ds + V(t+\Delta, x(t+\Delta)) \\
&\geq V_\Delta.
\end{aligned}$$

Since this holds with an arbitrary $\varepsilon > 0$, we have proved $V_\Delta \leq V(t,x)$.

Suppose $V_\Delta < V(t,x)$. Then there exists some $\hat{u} \in U^{[t,t+\Delta]}$ such that

$$\int_t^{t+\Delta} L(s,x(s),\hat{u}(s))ds + V(t+\Delta, x(t+\Delta)) < V(t,x).$$

Let

$$\varepsilon = V(t,x) - \left(\int_t^{t+\Delta} L(s,x(s),u(s))ds + V(t+\Delta t, x(t+\Delta t)) \right).$$

By the definition of $V(t+\Delta, x(t+\Delta))$, there exists some $u_\varepsilon \in U^{[t+\Delta,t_1]}$ such that

$$J(t+\Delta, x(t+\Delta t), u_\varepsilon) \leq V(t+\Delta t, x(t+\Delta t)) + \frac{\varepsilon}{2}.$$

Now, define $u \in U^{[t,t_1]}$ by

$$u(s) = \begin{cases} \hat{u}(s), & s \in [t, t+\Delta], \\ u_\varepsilon(s), & s \in (t+\Delta, t_1]. \end{cases}$$

We can verify that

$$J(t, x, u) = \int_t^{t_1} L(s, x(s), u(s))ds + K(t_1, x(t_1))$$

$$= \int_t^{t+\Delta} L(s, x(s), \hat{u}(s))ds + \int_{t+\Delta}^{t_1} L(s, x(s), u_\varepsilon(s))ds + K(t_1, x(t_1))$$

$$\leq \int_t^{t+\Delta} L(s, x(s), \hat{u}(s))ds + V(t+\Delta, x(t+\Delta)) + \frac{\varepsilon}{2}$$

$$= V(t, x) - \varepsilon + \frac{\varepsilon}{2} = V(t, x) - \frac{\varepsilon}{2}.$$

This contradicts the definition of $V(t, x)$. Hence, we must have $V_\Delta = V(t, x)$. ∎

2.2.2 Hamilton–Jacobi–Bellman Equation

From the principle of optimality, we can derive the celebrated Hamilton–Jacobi–Bellman (HJB) equation for the value function.

By (2.4), we obtain

$$0 = \inf_{u \in U^{[t,t+\Delta]}} \left\{ \frac{1}{\Delta} \int_t^{t+\Delta} L(s, x(s), u(s))ds + \frac{V(t+\Delta, x(t+\Delta)) - V(t, x)}{\Delta} \right\},$$

Letting $\Delta \to 0$ gives

$$0 = \inf_{u \in U} \left\{ L(t, x, u) + \frac{\partial V}{\partial t}(t, x) + \frac{\partial V}{\partial x}(t, x)f(t, x, u) \right\}.$$

Since $\frac{\partial V}{\partial t}(t, x)$ does not depend on u, we can pull it out of the infimum and write the equation as follows:

$$-\frac{\partial V}{\partial t}(t, x) = \inf_{u \in U} \left\{ L(t, x, u) + \frac{\partial V}{\partial x}(t, x)f(t, x, u) \right\}. \tag{2.5}$$

This equation is called the **HJB equation**. Furthermore, by the definition of the value function, it also satisfies the boundary condition:

$$V(t_1, x) = K(t_1, x). \tag{2.6}$$

Assuming sufficient smoothness of the value function, the above derivation shows that the value function necessarily satisfies the HJB equation together with the above boundary condition.

2.2.3 A Sufficient Condition for Optimality

The next result shows how the HJB equation can provide a sufficient condition for optimality.

Proposition 2.2 *(Sufficient condition for optimality by the HJB equation)*
Suppose that there exists a continuously differentiable function $\hat{V}$: $[t_0, t_1] \times \mathbb{R}^n \to \mathbb{R}$ that satisfies the HJB equation (2.5) with the boundary condition (2.6). Suppose further that there exists a control $\hat{u}$: $[t_0, t_1] \to U$ such that $\hat{u}$ and the corresponding state trajectory $\hat{x}$ with $x(t_0) = x_0$ satisfy

$$L(t, \hat{x}(t), \hat{u}(t)) + \frac{\partial \hat{V}}{\partial x}(t, \hat{x}(t))f(t, \hat{x}(t), \hat{u}(t))$$

$$= \min_{u \in U} \left\{ L(t, \hat{x}(t), u) + \frac{\partial \hat{V}}{\partial x}(t, \hat{x}(t))f(t, \hat{x}(t), u) \right\}, \quad t \in [t_0, t_1]. \quad (2.7)$$

Then $\hat{V}(t_0, x_0)$ is the optimal cost, i.e. $\hat{V}(t_0, x_0) = V(t_0, x_0)$, where V is the value function, and $\hat{u}$ is an optimal control.

Proof: By the HJB equation (2.5) for $\hat{V}$ with $x = \hat{x}(t)$ and using (2.7), we have

$$-\frac{\partial \hat{V}}{\partial t}(t, \hat{x}(t)) = \inf_{u \in U} \left\{ L(t, \hat{x}(t), u) + \frac{\partial V}{\partial x}(t, \hat{x}(t))f(t, \hat{x}(t), u) \right\}$$

$$= \min_{u \in U} \left\{ L(t, \hat{x}(t), u) + \frac{\partial \hat{V}}{\partial x}(t, \hat{x}(t))f(t, \hat{x}(t), u) \right\}$$

$$= L(t, \hat{x}(t), \hat{u}(t)) + \frac{\partial \hat{V}}{\partial x}(t, \hat{x}(t))f(t, \hat{x}(t), \hat{u}(t)),$$

which implies

$$0 = L(t, \hat{x}(t), \hat{u}(t)) + \frac{d}{dt}\hat{V}(t, \hat{x}(t)), \quad t \in [t_0, t_1].$$

Integrating this from t_0 to t_1 gives

$$0 = \int_{t_0}^{t_1} L(t, \hat{x}(t), \hat{u}(t))dt + \hat{V}(t_1, \hat{x}(t_1)) - \hat{V}(t_0, \hat{x}(t_0))$$

$$= \int_{t_0}^{t_1} L(t, \hat{x}(t), \hat{u}(t))dt + K(\hat{x}(t_1)) - \hat{V}(t_0, x_0),$$

which implies

$$\hat{V}(t_0, x_0) = \int_{t_0}^{t_1} L(t, \hat{x}(t), \hat{u}(t))dt + K(\hat{x}(t_1)) = J(t_0, x_0, \hat{u}). \quad (2.8)$$

Now, let x be any trajectory that corresponds to some u and the initial condition $x(t_0) = x_0$. The HJB equation (2.5) for $\hat{V}$ with $x = x(t)$ gives

$$-\frac{\partial \hat{V}}{\partial t}(t, x(t)) = \inf_{u \in U} \left\{ L(t, x(t), u) + \frac{\partial V}{\partial x}(t, x(t))f(t, x(t), u) \right\}$$

$$\leq L(t, x(t), u(t)) + \frac{\partial \hat{V}}{\partial x}(t, x(t))f(t, x(t), u(t)).$$

Repeating the above derivation, we obtain

$$0 \leq \int_{t_0}^{t_1} L(t, x(t), u(t))dt + K(x(t_1)) - \hat{V}(t_0, x_0),$$

which implies

$$\hat{V}(t_0, x_0) \leq J(t_0, x_0, u). \tag{2.9}$$

In view of (2.8) and (2.9), we conclude that $\hat{V}(t_0, x_0)$ is the optimal cost and $\hat{u}$ is an optimal control. ∎

2.2.4 Infinite-Horizon Problems

So far we have only considered the finite-horizon cost (2.2). In this subsection, we consider an infinite-horizon optimal control problem subject to

$$\dot{x} = f(x, u), \quad x(0) = x_0, \tag{2.10}$$

and the cost

$$J(x_0, u) = \int_0^\infty L(x, u)dt. \tag{2.11}$$

Compared with (2.1) and (2.2), the right-hand side f and running cost L here are assumed to be time invariant. Under this assumption, the cost and hence the value function do not depend on the initial time. This justifies our choice of 0 as the initial time.

Since the value function is independent of t, it takes the form $V = V(x)$ and the HJB equation reduces to

$$0 = \inf_{u \in U} \left\{ L(x, u) + \frac{\partial V}{\partial x}(x)f(x, u) \right\}. \tag{2.12}$$

Introduce the *Hamiltonian*

$$H(x, u, p) := -L(x, u) + p^T f(x, u), \tag{2.13}$$

where $p \in \mathbb{R}^n$.

The following theorem provides a sufficient condition for verifying a feedback controller to be both optimal and asymptotically stabilizing. In a sense, it connects the sufficient conditions for finite-horizon optimality (Proposition 2.2) with stability analysis using Lyapunov functions (Theorem 1.1). We are only interested in controllers that are stabilizing. To this end, denote by $x(t; x_0, u)$ the solution of (2.10) under an input $u(\cdot)$ and define the set of stabilizing controllers for (2.10) w.r.t. to a compact set $A \subseteq \mathbb{R}^n$ as

$$\mathcal{K}(x_0) := \left\{ u \in U^{[0,\infty)} : \|x(t; x_0, u)\|_A \to 0 \text{ as } t \to \infty \right\}.$$

Theorem 2.1 *(Optimal feedback control)* Let A be a compact invariant set of system (2.10). Let $D \subseteq \mathbb{R}^n$ be an open set containing A, $V : D \to \mathbb{R}$ be a continuously differentiable function, and $\kappa : D \to U$ be locally Lipschitz. Assume that V is positive definite with respect to A and

$$\frac{dV}{dx}(x)f(x, \kappa(x))$$

is negative definite with respect to A. Furthermore,

$$H\left(x, \kappa(x), -\left(\frac{dV}{dx}\right)^T \right) = 0, \quad \forall x \in D, \tag{2.14}$$

and

$$H\left(x, u, -\left(\frac{dV}{dx}\right)^T \right) \leq 0, \quad \forall x \in D, \quad \forall u \in U. \tag{2.15}$$

Then A is Uniformly Asymptotically Stable (UAS) for the closed-loop system with (2.10) and $u(\cdot) = \kappa(x(\cdot))$. Furthermore,

$$J(x_0, \kappa(x(\cdot))) = V(x_0) = \min_{u(\cdot) \in \mathcal{K}(x_0)} J(x_0, u(\cdot)). \tag{2.16}$$

If the above conditions hold for $D = \mathbb{R}^n$ and V is radially unbounded, then A is Globally Uniformly Asymptotically Stable (GUAS) for the closed-loop system.

Proof: Uniform asymptotic stability of A under the feedback controller $u = \kappa(x)$ follows directly from Theorem 1.1.

By the definition of H, Eq. (2.14) is

$$L(x, \kappa(x)) + \frac{dV}{dx}f(x, \kappa(x)) = 0, \tag{2.17}$$

and the inequality (2.15) is

$$L(x, u) + \frac{dV}{dx}f(x, u) \geq 0. \tag{2.18}$$

By (2.17), we have

$$\begin{aligned}
J(x_0, \kappa(x(\cdot))) &= \int_0^\infty L(x(t), \kappa(x(t)))dt \\
&= -\int_0^\infty \frac{dV}{dx}(x(t))f(x(t), \kappa(x(t)))dt \\
&= -\int_0^\infty \dot{V}(x(t))dt \\
&= V(x(0)) - \lim_{t \to \infty} V(x(t)) = V(x_0), \tag{2.19}
\end{aligned}$$

where $\lim_{t \to \infty} V(x(t)) = 0$ because $\|x(t)\|_A \to 0$ as $t \to \infty$ and $V = 0$ on A.

Let $u(\cdot)$ be any control that is stabilizing for (2.10) w.r.t. the set A. By (2.18) and similar to the derivation above, we can obtain

$$
\begin{aligned}
J(x_0, u(\cdot)) &= \int_0^\infty L(x(t), u(t)) dt \\
&\geq - \int_0^\infty \frac{dV}{dx}(x(t)) f(x(t), u(t)) dt \\
&= - \int_0^\infty \dot{V}(x(t)) dt \\
&= V(x(0)) - \lim_{t \to \infty} V(x(t)) = V(x_0).
\end{aligned} \tag{2.20}
$$

We have proved (2.16). ∎

Remark 2.1 We wrote (2.14) and (2.15) in terms of Hamiltonian to emphasize that

$$
0 = H\left(x, \kappa(x), -\left(\frac{dV}{dx}\right)^T\right) \geq H\left(x, u, -\left(\frac{dV}{dx}\right)^T\right), \quad \forall x \in D, \quad \forall x \in U.
$$

This inequality shows that the optimal control maximizes the Hamiltonian, a necessary condition for optimality from the celebrated *Pontryagin's maximum principle*. On the other hand, if we explicitly write the above condition out, we obtain

$$
\begin{aligned}
L(x, \kappa(x)) &+ \frac{\partial V}{\partial x}(x) f(x, \kappa(x)) \\
&= \min_{u \in U} \left\{ L(x, u) + \frac{\partial V}{\partial x}(x) f(x, u) \right\}, \quad \forall x \in D.
\end{aligned} \tag{2.21}
$$

This resembles condition (2.7) in Proposition 2.2 (sufficient conditions for optimality), but now stated for a feedback controller. One can easily see that the same argument is behind both proofs.

2.3 Linear Quadratic Regulator

2.3.1 Differential Riccati Equation

In general, solving the partial differential HJB equation (2.5) is a difficult task. We now consider a special form of optimal control problems that is amenable to more efficient solutions.

Consider a Linear Time-Invariant (LTI) system of the form

$$
\dot{x} = Ax + Bu, \quad x(t_0) = x_0, \tag{2.22}
$$

where $x \in \mathbb{R}^n$ is the state and $u \in \mathbb{R}^m$ is the control, along with a cost functional defined by

$$
J(x_0, u) = \int_{t_0}^{t_1} [x^T(t) Q x(t) + u^T(t) R u(t)] dt + x^T(t_1) P_1 x(t_1), \tag{2.23}
$$

where $Q \in \mathbb{R}^{n \times n}$ and $P_1 \in \mathbb{R}^{n \times n}$ are symmetric and positive semidefinite (denoted by $Q \geq 0$ and $P_1 \geq 0$) and $R \in \mathbb{R}^{m \times m}$ is symmetric and positive definite (denoted by $R > 0$). This is clearly a special case of the general optimal control problem defined by the dynamics (2.1) and cost (2.2) with

$$f(t, x, u) = Ax + Bu, \quad L(t, x, u) = x^T Q x + u^T R u, \quad K(t, x) = x^T P_1 x, \quad U = \mathbb{R}^k.$$

This problem is known as the (finite-horizon) **Linear Quadratic Regulator** (**LQR**) problem. For this problem, the HJB equation for the optimal value function becomes

$$-\frac{\partial V}{\partial t}(t, x) = \inf_{u \in U} \left\{ x^T Q x + u^T R u + \frac{\partial V}{\partial x}(t, x)(Ax + Bu) \right\}.$$

Within the bracket is a quadratic function of u. We can find (by computing the critical point) that

$$u = -\frac{1}{2} R^{-1} B^T \left(\frac{\partial V}{\partial x}(t, x) \right)^T \tag{2.24}$$

minimizes the quadratic function in the bracket. With this, the HJB equation simplifies to

$$-\frac{\partial V}{\partial t}(t, x) = x^T Q x + \frac{\partial V}{\partial x}(t, x)Ax - \frac{1}{4}\frac{\partial V}{\partial x}(t, x)BR^{-1}B^T \left(\frac{\partial V}{\partial x}(t, x) \right)^T, \tag{2.25}$$

with the boundary condition

$$V(t_1, x) = x^T P_1 x. \tag{2.26}$$

This suggests a "guess" of a value function of the form

$$V(t, x) = x^T P(t) x, \tag{2.27}$$

where $P(t)$ is assumed to be a symmetric (which we will justify later) and continuously differentiable matrix function that satisfies the boundary condition $P(t_1) = P_1$. Then the HJB equation becomes

$$-x^T \dot{P}(t)x = x^T Q x + 2x^T P(t)Ax - x^T P(t)BR^{-1}B^T P(t)x,$$

which can be rewritten as follows·

$$x^T(\dot{P}(t) + Q + P(t)A + A^T P(t) - P(t)BR^{-1}B^T P(t))x = 0.$$

Clearly, if $P(t)$ is a solution to the matrix differential equation

$$\dot{P}(t) = -Q - P(t)A - A^T P(t) + P(t)BR^{-1}B^T P(t), \tag{2.28}$$

with the condition $P(t_1) = P_1$, then the above HJB equation, together with its boundary condition, is satisfied with V defined by (2.27). By Proposition 2.2, $V(t, x) = x^T P(t)x$ is indeed the value function and, by (2.24), an optimal control is given by

$$u(t) = -\frac{1}{2} R^{-1} B^T \left(\frac{\partial V}{\partial x}(t, x(t)) \right)^T = -R^{-1}B^T P(t)x(t). \tag{2.29}$$

To find the value function and the resulting optimal control, we only need to solve the matrix differential equation (2.28), which is the well-known (quadratic) *Differential Riccati Equation (DRE)*.

We also show an alternative, and more elementary, derivation of a solution to the finite-horizon LQR problem using the DRE. The technique is essentially "completing the square" as we do to minimize a quadratic function defined on $\mathbb{R}^n$.

Let $P(t) \in \mathbb{R}^{n \times m}$ be a differentiable matrix function to be determined. Suppose that $P(t)$ is symmetric for all t and satisfies a boundary condition $P(t_1) = P_1$. Write

$$J(x_0, u) = \int_{t_0}^{t_1} [x^T Q x + u^T R u] dt + x^T(t_1) M x(t_1)$$

$$= x_0^T P(t_0) x_0 + x^T(t_1) P(t_1) x(t_1) - x_0^T P(t_0) x_0 + \int_{t_0}^{t_1} [x^T Q x + u^T R u] dt$$

$$= x_0^T P(t_0) x_0 + x^T(t) P(t) x(t) \big|_{t=t_0}^{t=t_1} + \int_{t_0}^{t_1} [x^T Q x + u^T R u] dt$$

$$= x_0^T P(t_0) x_0 + \int_{t_0}^{t_1} \frac{d}{dt} \left[x(t)^T P(t) x(t) \right] dt + \int_{t_0}^{t_1} [x^T Q x + u^T R u] dt$$

$$= x_0^T P(t_0) x_0$$
$$+ \int_{t_0}^{t_1} [(Ax + Bu)^T P(t) x + x^T P(Ax + Bu) + x^T \dot{P}(t) x + x^T Q x + u^T R u] dt$$

$$= x_0^T P(t_0) x_0$$
$$+ \int_{t_0}^{t_1} x(t)^T [A^T P(t) + P(t)A + \dot{P}(t) + Q] x(t) + u^T R u + 2u^T B^T P(t) x \, dt$$

$$= x_0^T P(t_0) x_0 + \int_{t_0}^{t_1} x(t)^T [A^T P(t) + P(t)A + \dot{P}(t) + Q] x(t) dt$$
$$+ \int_{t_0}^{t_1} (u - Kx)^T R(u - Kx) - (Kx)^T R(Kx)$$

$$= x_0^T P(t_0) x_0$$
$$+ \int_{t_0}^{t_1} x(t)^T [A^T P(t) + P(t)A + \dot{P}(t) + Q - P(t) B R^{-1} B^T P(t)] x(t) dt$$

$$+ \int_{t_0}^{t_1} (u - Kx)^T R(u - Kx) dt, \tag{2.30}$$

where $K(t) = -R^{-1} B^T P(t)$ is obtained by completing the square

$$u^T R u + 2u^T B^T P(t) x = (u - Kx)^T R(u - Kx) - (Kx)^T R(Kx),$$

and we used the fact that P is symmetric and R is positive definite (and hence also invertible).

Now, suppose that $P(t)$ satisfies DRE (2.28) subject to the boundary condition $P(t_1) = P_1$. Then the above Eq. (2.30) for J becomes

$$J(x_0, u) = x_0^T P(t_0) x_0 + \int_{t_0}^{t_1} (u - Kx)^T R(u - Kx) dt. \tag{2.31}$$

Since R is positive definite, we have $(u - Kx)^T R(u - Kx) \geq 0$. This cost is clearly minimized if we choose

$$u(t) = K(t)x(t) = -R^{-1} B^T P(t)x(t). \tag{2.32}$$

This agrees with the optimal control (2.29) we obtained using the HJB approach. The resulting optimal cost is given by $x_0^T P(t_0) x_0$. Clearly, (t_0, x_0) can in fact be replaced with (t, x), where $t < t_1$ and $x \in \mathbb{R}^n$, and the argument above shows that the value function is

$$V(t, x) = x^T P(t)x,$$

which agrees with (2.27).

The following result summarizes what we have derived so far.

Proposition 2.3 *Suppose that there exists a symmetric and differentiable matrix function $P : [t_0, t_1] \to \mathbb{R}^{n \times n}$ satisfies the DRE (2.28) with the boundary condition $P(t_1) = P_1$. Then the LQR problem (2.22) and (2.23) has the optimal cost given by $J^* = x_0^T P(t_0) x_0$, which is achieved by the optimal state-feedback control $u^*(t) = -R^{-1} B^T P(t)x(t)$.*

A natural question to ask now is how to ensure that a solution to the DRE (2.28) exists, not just on a small time interval containing t_1 but also can be extended backward to cover any initial time t_0 of interest. Clearly, the matrix differential equation (2.28) can be written as an Ordinary Differential Equation (ODE) of dimension n^2. The right-hand side of this ODE is quadratic in its state variables; hence, it is continuously differentiable and locally Lipschitz. By Remark 1.1, with respect to the terminal condition $P(t_1) = M$, there exists a unique local solution $P(t)$ defined on an interval of the form $[t_1 - c, t_1 + c]$, $c > 0$. By Remark 1.2, if we can show that $P(t)$ remains bounded on each bounded interval of the form $(\alpha, t_1]$, it is guaranteed that there is a unique solution $P(t)$ to the DRE (2.28) defined for all $t < t_1$. In other words, solutions to (2.28) are always backward complete. We establish such properties in the next proposition.

Proposition 2.4 *Let $P(t)$ be a solution to the DRE (2.28), subject to the boundary condition $P(t_1) = P_1$, where P_1 is symmetric and positive semidefinite, defined on $(\alpha, t_1]$ for some $\alpha < t_1$. Then the following hold:*

1. $P(t)$ is symmetric, i.e. $P(t)^T = P(t)$, for all $t \in (\alpha, t_1]$.

2. $P(t) \geq 0$ *(i.e. positive semidefinite) for all* $t \in (\alpha, t_1]$.
3. $P(t)$ *is bounded on* $(\alpha, t_1]$.

Consequently, $P(t)$ is defined for all $t < t_1$.

Proof:
1. If $P(t)$ satisfies (2.28), it can be easily verified that $P(t)^T$ also satisfies (2.28). It also satisfies the boundary condition $P(t_1) = P_1$ because P_1 is symmetric. By the uniqueness of solutions for (2.28) (Remark 1.1), we have $P(t)^T = P(t)$ for all t.
2. Note that the derivation that led to (2.30) and (2.31) is valid with t_0 replaced by any $t < t_1$, which would lead to

$$J(x, u) = \int_t^{t_1} [x^T(s)Qx(s) + u^T(s)Ru(s)]ds + x^T(t_1)P_1x(t_1)$$

$$= x(t)^T P(t)x(t) + \int_t^{t_1} (u(s) - K(s)x(s))^T R(u(s) - K(s)x(s))dt.$$

By the fact that Q and P_1 are positive semidefinite and R is positive definite, we have $J \geq 0$. Fix t, we can choose $u(s) = K(s)x(s)$ for $s \in [t, t_1]$ to minimize the cost. The optimal cost (denoted by J^*) is still nonnegative by definition. However, this implies

$$J^* = x(t)^T P(t)x(t) \geq 0.$$

Since this would hold for an arbitrary $x(t) \in \mathbb{R}^n$, we must have $P(t) \geq 0$.
3. Starting from any $t < t_1$, we consider the cost of controlling the LTI system from an initial state $x(t)$ on $[t, t_1]$. The above argument shows that the optimal cost is

$$J^* = x(t)^T P(t)x(t).$$

In particular, this cost will be no larger than the cost of applying the zero control on $[t, t_1]$, which shows

$$x(t)^T P(t)x(t) \leq \int_t^{t_1} x^T(s)Qx(s)ds + x^T(t_1)P_1x(t_1),$$

where $x(s) = e^{A(s-t)}x(t)$ for all $s \in [t, t_1]$. For any $t \in (\alpha, t_1]$, if we choose $x(t)$ from a bounded set (e.g. the closed unit Euclidean ball), then there exists a uniform bound C (that depends on $(\alpha, t_1]$ as well as A and P_1) for the right-hand side of the above inequality. In parts (1) and (2), we have shown that $P(t)$ is symmetric and positive semidefinite. An equivalent condition for a matrix to be positive semidefinite is that all of its principal minors are nonnegative. For the sake of contradiction, assume that $P(t)$ becomes unbounded as $t \to \alpha^-$ (this is the only possibility because $P(t)$ is defined and continuous on $(\alpha, t_1]$). First, assume that all entries on the diagonal of $P(t)$ are bounded as $t \to \alpha^-$ and there exists one term $P_{ij}(t)$ $(i \neq j)$ is unbounded as $t \to \alpha^-$. Note that we have $P_{ij}(t) = P_{ji}(t)$ due to symmetry. The determinant of this principal matrix

$$\begin{bmatrix} P_{ii}(t) & P_{ij}(t) \\ P_{ij}(t) & P_{jj}(t) \end{bmatrix}$$

will become negative as $t \to \alpha^-$, which contradicts that $P(t)$ is positive semidefinite. Now, suppose $P_{ii}(t)$ is unbounded for some i as $t \to \alpha^-$. Pick $x(t) = e_i$, which is a vector with the ith component equal to 1 and others equal to 0. We obtain $x(t)^T P(t)x(t) = e_i^T P(t)e_i = P_{ii}(t) \to \infty$ as $t \to \alpha^-$. This contradicts that $x(t)^T P(t)x(t)$ remains bounded for $x(t)$ taken from a bounded set.

By the continuation of solutions (Remark 1.2), we obtain that $P(t)$ is well defined for all $t < t_1$. ∎

Example 2.1 *The condition that $P_1 \geq 0$ is essential for ensuring that $P(t)$ is defined for all $t < t_1$. Consider a scalar example where $A = 0$, $B = 1$, $Q = 1$, $R = 1$. Then DRE is $\dot{P} = P^2 - 1$. If $P(1) = -2$, then one can solve it directly to obtain $P(t) = \frac{3e^{2t}+e^2}{e^2-3e^{2t}}$. Clearly, $P(t)$ blows up as $t \to (1 - \ln 3/2)^+$.*

2.3.2 Algebraic Riccati Equation

We now consider an infinite-horizon LQR problem for the LTI system (2.22) with the

$$J(x_0, u) = \int_0^\infty [x^T(t)Qx(t) + u^T(t)Ru(t)]dt, \tag{2.33}$$

where $Q \geq 0$ and $R > 0$.

Similar to the finite-horizon LQR, we show two different ways of deriving the optimal controller. First, the HJB equation (2.12) becomes

$$0 = \inf_{u \in U} \left\{ x^T Qx + u^T Ru + \frac{\partial V}{\partial x}(x)(Ax + Bu) \right\}.$$

Note that V is independent of t due to the choice of cost and dynamics (see Section 2.2.4). We can minimize the right-hand side (which is quadratic in u) by choosing

$$u = -\frac{1}{2}R^{-1}B^T \left(\frac{\partial V}{\partial x}(x) \right)^T. \tag{2.34}$$

With this, the HJB equation reduces to

$$0 = x^T Qx + \frac{\partial V}{\partial x}(x)Ax - \frac{1}{4}\frac{\partial V}{\partial x}(x)BR^{-1}B^T \left(\frac{\partial V}{\partial x}(x) \right)^T. \tag{2.35}$$

If we guess the value function V to be the form

$$V(x) = x^T Px, \tag{2.36}$$

where $P \in \mathbb{R}^{n \times n}$ is a symmetric matrix, then the HJB equation becomes a purely algebraic relation

$$x^T(Q + PA + A^TP - PBR^{-1}B^TP)x = 0.$$

Clearly, this holds if P solves the algebraic matrix equation

$$0 = Q + PA + A^TP - PBR^{-1}B^TP. \tag{2.37}$$

With P, (2.34) gives a feedback controller of the form

$$u = -R^{-1}B^TPx. \tag{2.38}$$

Equation (2.37) is known as the *Algebraic Riccati Equation (ARE)*.

Optimality and stability: Suppose that (2.37) does have a solution. To show that (2.38) indeed gives an optimal controller, we can run the proof of optimality as in Theorem 2.1. For the argument to go through, however, we need $V(x(t)) \to 0$ as $t \to \infty$ (see Eqs. (2.19) and (2.20)). On the other hand, stability is a fundamental requirement for a dynamical system. Hence, we should enforce $x(t) \to 0$ as $t \to \infty$, which in term implies $V(x(t)) \to 0$ as $t \to \infty$. Asymptotic stability of the closed-loop system with the state-feedback controller (2.38) is equivalent to $A - BR^{-1}B^TP$ being Hurwitz, i.e. all of its eigenvalues have negative real parts. In other words, if there exists a symmetric matrix P that solves the ARE (2.37) and renders $A - BR^{-1}B^TP$ Hurwitz, then stability and optimality of the closed-loop system are guaranteed.

We summarize this result into the next proposition and prove it directly using an elementary approach. The proof is essentially the infinite-horizon version of the alternative proof we did for finite-horizon LQR in Section 2.3.1.

Proposition 2.5 *If there exists a symmetric matrix P such that the ARE (2.37) holds and $A - BR^{-1}B^TP$ is Hurwitz, then the state-feedback controller (2.38) renders the closed-loop system asymptotically stable and is optimal among all stabilizing controllers.*

Proof: Consider any u such that $\lim_{t \to \infty} x(t) = 0$. With P, we can write

$$
\begin{aligned}
J(x_0, u) &= \int_0^\infty [x^TQx + u^TRu]dt \\
&= x_0^TPx_0 + 0 - x_0^TPx_0 + \int_0^\infty [x^TQx + u^TRu]dt \\
&= x_0^TP(t_0)x_0 + x^T(t)Px(t)|_{t=0}^{t=\infty} + \int_0^\infty [x^TQx + u^TRu]dt \\
&= x_0^TPx_0 + \int_0^\infty \frac{d}{dt}\left[x(t)^TPx(t)\right]dt + \int_0^\infty [x^TQx + u^TRu]dt
\end{aligned}
$$

$$= x_0^T Px_0 + \int_0^\infty [(Ax + Bu)^T Px + x^T P(Ax + Bu) + x^T Qx + u^T Ru]dt$$

$$= x_0^T Px_0 + \int_0^\infty x(t)^T [A^T P + PA + Q]x(t) + u^T Ru + 2u^T B^T Pxdt$$

$$= x_0^T Px_0 + \int_0^\infty x(t)^T [A^T P + PA + Q]x(t)dt$$

$$+ \int_0^\infty (u - Kx)^T R(u - Kx) - (Kx)^T R(Kx)$$

$$= x_0^T P(t_0)x_0 + \int_0^\infty x(t)^T [A^T P + PA + Q - PBR^{-1}B^T P]x(t)dt$$

$$+ \int_0^\infty (u - Kx)^T R(u - Kx)dt$$

$$= x_0^T P(t_0)x_0 + \int_0^\infty (u - Kx)^T R(u - Kx)dt, \tag{2.39}$$

where $K = -R^{-1}B^T P$. We used the fact that P solves the ARE (2.37) to obtain the last equality and $\lim_{t\to\infty} x(t) = 0$ to obtain the third one. The rest of steps are algebraic manipulation. Since R is positive definite, setting $u = Kx = -R^{-1}B^T P$ would minimize $J(x_0, u)$, while $u = Kx$ remains stabilizing because $A + BK = A - BR^{-1}B^T P$ is Hurwitz. ∎

The ARE can have multiple solutions.

Example 2.2 *Consider the system $\dot{x} = x + u$ with a cost functional defined by $J(x_0, u) = \int_0^\infty u^2(t)dt$ (i.e. $A = 1$, $B = 1$, $Q = 0$, and $R = 1$). The ARE (2.37) reduces to $2P - P^2 = 0$. The solution $P = 0$ gives $u = 0$, which is not stabilizing. The solution $P = 2$ gives $u = -2x$. The closed-loop system is $\dot{x} = -x$, which gives $x(t) = e^{-t}x_0$ and $u(t) = -2x(t) = -2e^{-t}x_0$. The cost is verified to be $J(x_0, u) = \int_0^\infty u^2(t)dt = \int_0^\infty 4e^{-2t}|x_0|^2 dt = 2x_0^2$. This is consistent with Proposition 2.5. This cost is optimal among all stabilizing controllers.*

Remark 2.2 *(Characterization of the existence of a stabilizing solution to ARE (2.37))* The following facts regarding the solutions of ARE (2.37) are stated without proof[2]:

1. A stabilizing solution to (2.37), if exists, is symmetric, unique, and positive semidefinite.
2. A stabilizing solution to (2.37) exists and (2.37) has a unique symmetric positive semidefinite solution (which in this case must be the stabilizing one) if and only if (A, B) is *stabilizable*, i.e. there exists a matrix K such that $A + BK$ is Hurwitz, and (A, Q) is *detectable* (which is equivalent to (A^T, Q^T) being stabilizable).

---

2 The readers are referred to Hespanha [2018] as a reference for proofs of these facts.

3. A stabilizing solution to (2.37) exists if and only if (A, B) is stabilizable and the *Hamiltonian matrix*

$$M = \begin{bmatrix} A & -BR^{-1}B^T \\ -Q & -A^T \end{bmatrix}$$

has no eigenvalues on the imaginary axis.

2.3.3 Convergence of Solutions to the Differential Riccati Equation

In Section 2.3.1, we have shown that the DRE characterizes the optimal value function for the finite-horizon LQR problem. More specifically, let $P(t) = P(t; t_1, P_1)$ denote the solution to the DRE (2.28) with the terminal condition $P(t_1) = P_1$, where $P_1 \geq 0$. The value function is given by $V(t, x) = x^T P(t)x$. Proposition 2.4 establishes that $P(t)$ is defined for all $t < t_1$. The resulting optimal feedback controller is provided by $u(t) = -R^{-1}B^T P(t)x(t)$.

In contrast, the infinite-horizon LQR problem is solved by the ARE (2.37). By Remark 2.2, the ARE (2.37) has a unique symmetric positive semidefinite solution that is also stabilizing, provided that (A, B) is stabilizable and (A, Q) is detectable. To distinguish with the time-varying solution $P(t)$ to DRE (2.28), we denote the unique stabilizing solution to ARE (2.37) by $\bar{P}$, which gives the time-invariant optimal value function as $V(x) = x^T \bar{P}x$ and the optimal feedback controller $u = -R^{-1}B^T \bar{P}x$.

A natural question to ask is how are $P(t)$ and $\bar{P}$ related? Before we discuss this, note that since the dynamics (2.22) and cost (2.23) are time invariant, we have $P(t; t_1, P_1) = P(t - t_1, 0, P_1)$. It is tempting to conjecture that as $t \to -\infty$ or $t_1 \to \infty$, the limit of $P(t; t_1, P_1)$ would approach $\bar{P}$.

This conjecture is not true in general, even if we know that $P(t; t_1, P_1)$ is defined for all $t < t_1$. The following example shows that $P(t)$ can remain oscillatory as $t - t_1 \to -\infty$.

Example 2.3 *Consider [Callier and Willems, 1981]*

$$A = \begin{bmatrix} 1 & 1 \\ -1 & 1 \end{bmatrix}, \ B = \begin{bmatrix} 1 & 0 \\ 0 & 1 \end{bmatrix}, \ R = \begin{bmatrix} 1 & 0 \\ 0 & 1 \end{bmatrix}, \ Q = \begin{bmatrix} 0 & 0 \\ 0 & 0 \end{bmatrix}, \ P_1 = \begin{bmatrix} 1 & 0 \\ 0 & 0 \end{bmatrix}.$$

Then it can be verified that

$$P(t; t_1, P_1) = \frac{2}{1 + e^{2(t-t_1)}} \begin{bmatrix} \cos^2(t_0 - t_1) & -\sin(t - t_1)\cos(t - t_1) \\ -\sin(t - t_1)\cos(t - t_1) & \sin^2(t - t_1) \end{bmatrix}.$$

Hence, $P(t; t_1, P_1)$ remains oscillatory as $t - t_1 \to -\infty$. For this example, (A, B) is controllable (and hence stabilizable). It is also easy to verify that the Hamiltonian matrix M defined in Remark 2.2 does not have any eigenvalues on the imaginary axis. Therefore, the ARE has a unique stabilizing solution, given by

$$\bar{P} = \begin{bmatrix} 2 & 0 \\ 0 & 2 \end{bmatrix}.$$

We can verify, however, (A, Q) is not detectable. By Remark 2.2, this means the ARE has more than one solution that is positive semidefinite. In fact, the zero matrix clearly solves the ARE for this problem since $Q = 0$.

If we assume that (A, B) is stabilizable and (A, Q) is detectable, then it turns out that $P(t; t_1, P_1)$ does converge to the unique stabilizing solution of ARE (2.37) as $t - t_1 \to \infty$. Since $P(t; t_1, P_1) = P(t - t_1, 0, P_1)$, it is without loss of generality to assume $t_1 = 0$ and denote the solution by $P(t)$. We are interested in the limit of $P(t)$ as $t \to -\infty$.

Proposition 2.6 *Suppose that (A, B) is stabilizable and (A, Q) is detectable. Let $\bar{P}$ denote the unique stabilizing solution of ARE (2.37). Let $P(t)$ denote the solution of the DRE (2.28) with $t_1 = 0$ and $P(0) = P_0 \geq 0$. We have*

$$P(t) = \bar{P} + e^{-\bar{A}^T t} \Theta(-t) e^{-\bar{A} t}, \quad \forall t \leq 0, \tag{2.40}$$

where $\bar{A} = A - BR^{-1}B^T\bar{P}$ is Hurwitz (because $\bar{P}$ is stabilizing) and

$$\Theta(-t) = (P_0 - \bar{P})\left[I + W(-t)(P_0 - \bar{P})\right]^{-1}, \tag{2.41}$$

is well defined for all $t \geq 0$. In (2.41), $W(\tau)$ is the closed-loop controllability grammian defined by

$$W(\tau) = \int_0^\tau e^{\bar{A}s} BR^{-1}B^T e^{\bar{A}^T s} ds, \quad \tau \geq 0.$$

Consequently, $P(t)$ converges to $\bar{P}$ exponentially, i.e. there exists some $\sigma > 0$ and some constant $c = c(P_0)$ such that

$$\left\|P(t) - \bar{P}\right\| \leq ce^{\sigma t}, \quad \forall t \leq 0. \tag{2.42}$$

The proof of Proposition 2.6 can be found in Callier et al. [1994]. For a simplified exposition when (A, B) is controllable and (A, Q) is observable, see Prach et al. [2015].

Let $\bar{P}$ be the stabilizing solution (2.37). With $\bar{A} = A - BR^{-1}B^T\bar{P}, S = BR^{-1}B^T$, and $\bar{Q} = \bar{P}S\bar{P} + Q \geq 0$, we can rewrite the ARE (2.37) as follows:

$$\bar{A}^T\bar{P} + \bar{P}\bar{A} = -\bar{Q} = -\bar{Q}^{\frac{1}{2}}\bar{Q}^{\frac{1}{2}}. \tag{2.43}$$

We have the following result.

Proposition 2.7 *Let $\bar{P}$ be the stabilizing solution (2.37). If (A, Q) is observable, then $\bar{P}$ is positive definite.*

Proof: It is well known that (A, Q) is observable if and only if $(A, Q^{1/2})$ is observable. It follows from (2.43) and the Lyapunov test for observability (see, e.g. [Hespanha, 2018, Theorem 15.10]) that $\bar{P}$ is positive definite. ∎

As a special case, if $Q > 0$, then $\bar{Q} > 0$. Note that Eq. (2.43) is the celebrated *Lyapunov equation*. Since $\bar{A}$ is Hurwitz, $\bar{P}$ is the unique solution to (2.43) and it is positive definite. Of course, $Q > 0$ also trivially implies that (A, Q) is observable and hence, the same conclusion of Proposition 2.7 follows.

2.3.4 Forward Propagation of the Differential Riccati Equation for Linear Quadratic Regulator

In Proposition 2.6, $P(t)$ ($t \leq 0$) is the backward solution of the DRE (2.28) with $P(0) = P_0$. Let $\hat{P}(t) = P(-t)$, where $t \geq 0$. Then

$$\dot{\hat{P}}(t) = -\dot{P}(-t) = P(-t)A + A^T P(-t) + Q - P(-t)BR^{-1}B^T P(-t)$$
$$= \hat{P}(t)A + A^T \hat{P}(t) + Q - \hat{P}(t)BR^{-1}B^T \hat{P}(t), \quad \hat{P}(0) = P_0.$$

In other words, $\hat{P}(t)$ solves the forward DRE:

$$\dot{P}(t) = P(t)A + A^T P(t) + Q - P(t)BR^{-1}B^T P(t), \quad P(0) = P_0. \tag{2.44}$$

Under the assumption that (A, B) is stabilizable and (A, Q) is detectable, Proposition 2.6 guarantees that the unique solution $P(t)$ of (2.44) on $[0, \infty)$ satisfies

$$\|P(t) - \bar{P}\| \leq ce^{-\sigma t}, \quad \forall t \geq 0. \tag{2.45}$$

Forward propagation of the DRE is more desirable when future dynamics are not necessarily known. This will be the premise of this book. By propagating a Riccati-like differential equation for a dynamical system that is incrementally learned, we seek to construct stabilizing controllers based on the solution of this forward DRE.

For the case of LQR control of LTI systems, $P(t)$ solved by forward propagating (2.44) defines a feedback controller of the form

$$u(t) = -R^{-1}B^T P(t)x(t), \quad t \geq 0. \tag{2.46}$$

Since $P(t) \to \bar{P}$ as $t \to \infty$, where $\bar{P}$ is the unique stabilizing solution of the ARE (2.37), it is reasonable to conjecture that (2.46) is also stabilizing. We prove that this is indeed the case in the following theorem.

Theorem 2.2 *Suppose that (A, B) is stabilizable and Q is positive definite. Let $P(t)$ denote the solution of the forward DRE (2.44) with $P(0) = P_0 \geq 0$. Consider the closed-loop system under the feedback controller (2.46):*

$$\dot{x} = f(t, x) := (A - BR^{-1}B^T P(t))x. \tag{2.47}$$

Then the origin is Globally Uniformly Exponentially Stable (GUES) for the closed-loop system.

Proof: Note that $Q > 0$ implies (A, Q) is observable (and hence detectable). By Proposition 2.6, $P(t)$ converges to the unique stabilizing solution $\bar{P}$ of the ARE (2.37). By Proposition 2.7, $\bar{P} > 0$. Consider the Lyapunov function

$$V(x) = x^T \bar{P} x.$$

Let $S = BR^{-1}B^T$. Clearly, S is a constant symmetric matrix and $S \geq 0$. Furthermore, Proposition 2.4 shows that $P(t)$ is also symmetric. We have

$$
\begin{aligned}
\dot{V}(x) &= x^T \bar{P}(A - SP(t))x + x^T \bar{P}(A^T - P(t)S)\bar{P}x \\
&= x^T \left[\bar{P}A + A^T\bar{P} - \bar{P}SP(t) - P(t)S\bar{P} \right] x \\
&= x^T \left[\bar{P}A + A^T\bar{P} + Q - \bar{P}S\bar{P} - Q + \bar{P}S\bar{P} - \bar{P}SP(t) - P(t)S\bar{P} \right] x \\
&= x^T \left[-Q - \bar{P}S\bar{P} + 2\bar{P}S\bar{P} - \bar{P}SP(t) - P(t)S\bar{P} \right] x \\
&\leq -x^T(Q - E(t))x,
\end{aligned}
$$

where $E(t) = 2\bar{P}S\bar{P} - \bar{P}SP(t) - P(t)S\bar{P}$, and we used the ARE (2.37) satisfied by $\bar{P}$ to obtain the last equality and $\bar{P}S\bar{P} \geq 0$ to obtain the inequality. Since $P(t) \to \bar{P}$ as $t \to \infty$, we have $E(t) \to 0$ as $t \to \infty$. This can be seen by taking the limit in $E(t) = \bar{P}S(\bar{P} - P(t)) + (\bar{P} - P(t))S\bar{P}$ as $t \to 0$. There exists some $T > 0$ and a positive constant μ such that $Q(t) - E(t) - \mu I \geq 0$. It follows that

$$\dot{V}(x) \leq -\mu\|x\|^2, \quad \forall t \geq T, \ \forall x \in \mathbb{R}^n. \tag{2.48}$$

Following a similar argument as in the proof for the Lyapunov theorem for exponential stability (see Theorem 1.2), we can show that there exists some $k > 0$ and $c > 0$ such that

$$\|x(t)\| \leq k \, \|x(T)\| \, e^{-c(t-T)}, \quad t \geq T. \tag{2.49}$$

By continuity and the fact that $P(t) \to \bar{P}$ as $t \to \infty$, $P(t)$ is bounded on $[0, \infty)$. It follows from (2.47) that there exist constants $C > 0$ and $M > 0$ such that

$$\|x(t)\| \leq M \, \|x_0\| \, e^{Ct}, \quad t \geq 0, \tag{2.50}$$

where $x_0 = x(0)$. Indeed, a simple comparison argument suffices to show this. We have

$$\frac{d}{dt}[\|x(t)\|^2] = 2x^T(t)(A - BR^{-1}B^T P(t))x(t) \leq 2C\|x(t)\|^2,$$

where such a constant $C > 0$ exists because $P(t)$ is bounded on $[0, \infty)$, which implies (2.50) with $M = 1$. On the interval $[0, T]$, we can rewrite (2.50) as follows:

$$\|x(t)\| \leq Me^{(C+c)t} \, \|x_0\| \, e^{-ct} \leq Me^{(C+c)T} \, \|x_0\| \, e^{-ct}, \quad \forall t \in [0, t]. \tag{2.51}$$

For $t \geq T$, rewrite (2.49) as follows:

$$\|x(t)\| \leq ke^{cT} \, \|x(T)\| \, e^{-ct} \leq kMe^{(C+c)T} \, \|x_0\| \, e^{-ct}, \quad \forall t \geq T, \tag{2.52}$$

where we used $x(T) \leq M \|x_0\| e^{CT}$ from (2.50). Combining (2.51) and (2.52) gives

$$\|x(t)\| \leq K \|x_0\| e^{-ct}, \quad \forall t \geq 0, \tag{2.53}$$

where $K = kMe^{(C+c)T}$ (note that $k \geq 1$ for (2.49) to hold). This verifies that the origin is GUES for the closed-loop system (2.47). ∎

2.4 Summary

This chapter gives a brief introduction to optimal control theory and its connection with closed-loop stability analysis. The emphasis here is on deriving optimal feedback control using the HJB approach. For infinite-horizon problems, Lyapunov analysis is used for asymptotic stability analysis of the closed-loop system under an optimal feedback control. For LTI systems, we give a brief account of the classic LQR theory, emphasizing the connections between the DRE and ARE as the control horizon approaches infinity.

The main references for writing this chapter include [Liberzon, 2003] for optimal control theory, Bernstein [1993] for Lyapunov analysis of optimal feedback control, and Callier and Willems [1981], Callier et al. [1994], and Prach et al. [2015] for convergence of DREs. Theorem 2.1 reformulates [Bernstein, 1993, Theorem 3.1] with regard to set stability. Theorem 2.2 strengthens the conclusion of [Prach et al., 2015, Theorem 3] (from asymptotic convergence to global uniform exponential stability) under slightly weaker conditions (stabilizability of (A, B) instead of controllability). The proof also differs with a simpler choice of time-invariant Lyapunov function (even though the time-varying one from [Prach et al., 2015, Theorem 3] also works). The idea of forward propagating the DRE will be revisited in later chapters for computing online optimal controllers using system models identified from sampled data.

Bibliography

Richard E. Bellman. *Dynamic Programming*. Princeton University Press, 1957.

Dennis S. Bernstein. Nonquadratic cost and nonlinear feedback control. *International Journal of Robust and Nonlinear Control*, 3(3):211–229, 1993.

Frank M. Callier and Jacques L. Willems. Criterion for the convergence of the solution of the Riccati differential equation. *IEEE Transactions on Automatic Control*, 26(6):1232–1242, 1981.

Frank M. Callier, Joseph Winkin, and Jacques L. Willems. Convergence of the time-invariant Riccati differential equation and LQ-problem: Mechanisms of attraction. *International Journal of Control*, 59(4):983–1000, 1994.

Joao P. Hespanha. *Linear Systems Theory*. Princeton University Press, 2018.

Daniel Liberzon. *Switching in Systems and Control*. Springer, 2003.

Daniel Liberzon. *Calculus of Variations and Optimal Control Theory*. Princeton University Press, 2011.

Anna Prach, Ozan Tekinalp, and Dennis S. Bernstein. Infinite-horizon linear-quadratic control by forward propagation of the differential Riccati equation. *IEEE Control Systems Magazine*, 35(2):78–93, 2015.

3

Reinforcement Learning

In this chapter, we introduce a popular computational framework for solving optimal control problems. While it is termed *Reinforcement Learning (RL)* [Sutton and Barto, 2018], the fundamental principle behind it is dynamic programming, the principle of optimality, and the celebrated Hamilton–Jacobi–Bellman (HJB) equation. Indeed, the traditional field of RL has now been intimately connected with the literature of optimal control for Markov Decision Processes (MDPs), which were pioneered by Richard Bellman in the 1950s [Bellman, 1957]. Nonetheless, the continuous control problems we are interested in solving are modeled by differential equations. The aim of this chapter is to provide an introduction to a subset of the standard RL techniques for solving this class of problems. Our focus is on continuous-time control systems, and we aim to provide self-contained proofs for all the technical results.

3.1 Control-Affine Systems with Quadratic Costs

Consider a class of optimal control problems with *control-affine* dynamics of the form

$$\dot{x} = f(x) + g(x)u, \tag{3.1}$$

where $f : \mathbb{R}^n \to \mathbb{R}^n$ and $g : \mathbb{R}^n \to \mathbb{R}^{n \times m}$, $x \in \mathbb{R}^n$ is the state, and u is the control input. The associated cost is quadratic in control and defined as follows:

$$J(x_0, u) = \int_0^\infty [Q(x(t)) + u^T(t)R(x(t))u(t)]dt, \tag{3.2}$$

Model-Based Reinforcement Learning: From Data to Continuous Actions with a Python-based Toolbox, First Edition. Milad Farsi and Jun Liu.

where $Q : \mathbb{R}^n \to \mathbb{R}$ is a positive definite function w.r.t. a compact set[1] $A \subseteq \mathbb{R}^n$ and $R : \mathbb{R}^n \to \mathbb{R}$ is symmetric and positive definite w.r.t. A. We assume that A is forward invariant for (3.1) when $u = 0$.

Let $V : \mathbb{R}^n \to \mathbb{R}$ denote the value function, i.e.

$$V(x) = \inf_{u \in (\mathbb{R}^n)^{[0,\infty)}} J(x, u). \tag{3.3}$$

We assume that the value function is continuously differentiable. For this problem, the HJB equation (2.12) becomes

$$0 = \inf_{u \in \mathbb{R}^m} \left\{ Q(x) + u^T R(x) u + \frac{\partial V}{\partial x}(x)(f(x) + g(x)u) \right\}.$$

We can minimize the right-hand side (which is quadratic in u) by choosing

$$u = \kappa(x) = -\frac{1}{2} R^{-1} g^T(x) \left(\frac{\partial V}{\partial x}(x) \right)^T. \tag{3.4}$$

With this, the HJB equation reduces to

$$0 = Q(x) + \frac{\partial V}{\partial x}(x) f(x) - \frac{1}{4} \frac{\partial V}{\partial x}(x) g(x) R^{-1}(x) g^T(x) \left(\frac{\partial V}{\partial x}(x) \right)^T. \tag{3.5}$$

Note that this derivation is almost identical to that for LQR, except for the nonlinear dependence of Q, R, f, and g in x, which does not contribute to the difficulty of this derivation because the key is that the cost is quadratic and the dynamics are linear in u.

We rephrase Theorem 2.1 for this problem as follows to provide a sufficient condition for optimal control and stabilization of (3.1).

Theorem 3.1 *Suppose that there exists a twice continuously differentiable function $V : \mathbb{R}^n \to \mathbb{R}$ that satisfies (3.5). Let $u = \kappa(x)$ be given by (3.4). Assume that V is positive definite with respect to A. Then A is Uniformly Asymptotically Stable (UAS) for the closed-loop system:*

$$\dot{x} = f(x) + g(x)\kappa(x), \tag{3.6}$$

and κ is optimal in the sense that

$$J(x_0, \kappa(x(\cdot))) = V(x_0) = \min_{u(\cdot) \in \mathcal{K}(x_0)} J(x_0, u(\cdot)), \tag{3.7}$$

where $\mathcal{K}(x_0)$ is the class of stabilizing controllers for (3.1) w.r.t. A. Furthermore, if V is radially unbounded, then A is Globally Uniformly Asymptotically Stable (GUAS) for (3.6).

---

1 While most of the literature focuses on optimal control and stabilization with the origin as the target set, the treatment of a compact target set A is not much more difficult in view of Theorem 2.1. Of course, when $A = \{0\}$, this reduces to the usual case.

Proof: This is an immediate corollary of Theorem 2.1. Indeed, (2.14) is simply the HJB equation (3.5) and (2.15) follows from

$$H\left(x,\kappa,-\left(\frac{dV}{dx}\right)^T\right) - H\left(x,u,-\left(\frac{dV}{dx}\right)^T\right) = -(u-\kappa(x))^T R(u-\kappa(x)) \leq 0.$$

The Lyapunov condition is verified by noting that V is positive definite and

$$\frac{\partial V}{\partial x}(f(x)+g(x)\kappa(x)) = -Q(x) - \frac{1}{4}\frac{\partial V}{\partial x}(x)g(x)R^{-1}(x)g^T(x)\left(\frac{\partial V}{\partial x}(x)\right)^T$$

is negative definite because $Q(x)$ is positive definite, all with respect to A. ∎

Remark 3.1 We require V to be twice continuously differentiable to ensure that κ is locally Lipschitz to be consistent with the regularity assumption we put on the right-hand side of the system. If V is only continuously differentiable, we can only ensure continuity of κ. In this case, uniqueness of solutions is no longer guaranteed, but stability analysis using Lyapunov functions still applies.

Theorem 3.1 shows that if we have access to the solution of the HJB equation 3.5, not only can we find an optimal controller but also we can also show that the controller is asymptotically stabilizing under some extra condition on V. Unfortunately, finding the solution to (3.5) is a difficult task. We have seen how linear dynamics and quadratic cost can lead to more efficient solutions by reducing the problem to solving Riccati equations. In general, finding the exact solution of (3.5) seems impossible. This has motivated the searching for approximate solutions of (3.5). In this regard, a few main considerations are in order

- Not only are we interested in finding approximate solutions of (3.5), but we would also like to be able to derive stabilizing controllers from these approximate value functions.
- We would like the approximate value functions to converge to the optimal value function and the derived controllers converge to an optimal controller.
- We would like to compute such approximate solutions without exact or any knowledge of the system dynamics (3.1), apart from the fact that they take the form in (3.1).

Achieving these will be the main tasks of the remaining sections of this chapter.

3.2 Exact Policy Iteration

Policy Iteration (PI) has its root in optimal control of MDPs [Bellman, 1957; Howard, 1960] (see also recent texts and monographs [Bertsekas, 2012, 2019]).

In this section, we introduce a basic form of PI for system (3.1) with cost (3.2). The algorithm dates back to Leake and Liu [1967], and its convergence was proved in Saridis and Lee [1979]. It is termed *exact policy iteration* because it uses precise knowledge of f and g in (3.1).

The exact PI algorithm starts with an initial policy $u = \kappa_0(x)$, which is assumed to be is a stabilizing controller. For each $i \geq 0$, Algorithm 3.1 (exact policy iteration) repeats the following two steps:

1. **(Policy evaluation)** Compute a value $V_i(x)$ for the policy κ_i by solving

$$Q(x) + \kappa_i^T(x)R(x)\kappa_i(x) + \frac{\partial V_i}{\partial x}(x)(f(x) + g(x)\kappa_i(x)) = 0 \qquad (3.8)$$

   subject to $V_i(x) = 0$ for all $x \in A$.

2. **(Policy improvement)** Update the policy

$$\kappa_{i+1}(x) = -\frac{1}{2}R^{-1}g^T(x)\left(\frac{\partial V_i}{\partial x}(x)\right)^T. \qquad (3.9)$$

Algorithm 3.1 Exact Policy Iteration

Require: f, g, κ_0

1: **repeat**
2:     Compute V_i such that (3.8) holds
3:     Update κ_{i+1} according to (3.9)
4:     $i = i + 1$
5: **until** $V_i = V_{i-1}$

Clearly, policy iteration can be seen as successive approximations to the solution of the HJB (3.5) and optimal controller (3.4). In its general form, Algorithm 3.1 is only of limited computational value for a few reasons. First, solving (3.8) is still difficult. In fact, it is at least as hard as finding a Lyapunov function for the nonlinear system (3.6) because V_i is indeed a Lyapunov function for system (3.6) with $\kappa = \kappa_i$. Second, strictly speaking, an algorithm needs to terminate. This successive approximation is unlikely to converge in a finite number of steps due to the continuous space. Despite these limitations, Algorithm 3.1 is of considerable theoretical value and can be shown to asymptotically converge to the optimal value V and optimal control κ. This is summarized in the following theorem. We require the following technical assumption.

Assumption 3.1 *We assume the following:*

1. *There exists a globally stabilizing policy κ_0 w.r.t. A such that solutions of the closed-loop system*

$$\dot{x} = f(x) + g(x)\kappa_0(x), \quad x(0) = x_0,$$

   satisfies $\|x(t)\|_A \to 0$ for all $x_0 \in \mathbb{R}^n$. We assume that κ_0 is locally Lipschitz.

2. The HJB equation (3.5) has a twice continuously differentiable solution $V : \mathbb{R}^n \to \mathbb{R}$ that is positive definite w.r.t. A and radially unbounded.

3. The sequence of functions $\{V_i\}$ returned by Algorithm 3.1 (exact policy iteration) is well defined, twice continuously differentiable on $\mathbb{R}^n$, and satisfy additionally that $\frac{\partial V_i}{\partial x_i}$ converges uniformly on every compact subset of $\mathbb{R}^n$.

Remark 3.2 Twice continuous differentiability of V and V_i can be replaced with continuous differentiability, if continuous differentiability (and local Lipschitz continuity) of κ is not required. Given V satisfying Assumption 3.1(2) and by Theorem 3.1, we know that $u = \kappa(x)$ defined by (3.4) renders A GUAS for the closed-loop system (3.6) and it is also optimal in the sense that

$$J(x_0, \kappa(x(\cdot))) = V(x_0) = \min_{u(\cdot) \in \mathcal{K}(x_0)} J(x_0, u(\cdot)), \tag{3.10}$$

where $\mathcal{K}(x_0)$ is the class of stabilizing controllers for (3.1) w.r.t. A. From this, we can immediately see that V satisfying Assumption 3.1(2) is unique. We denote this pair by V^* and κ^*.

Under Assumption 3.1, we show that $\{V_i\}$ and $\{\kappa_i\}$ returned by Algorithm 3.1 converge to V^* and κ^*, respectively, as $i \to \infty$.

Theorem 3.2 *Suppose that Assumption 3.1 holds. Then the sequences of $\{V_i\}$ and $\{\kappa_i\}$ returned by Algorithm 3.1 (exact policy iteration) satisfy:*

1. *Each κ_i renders the set A GUAS for the closed-loop system*

$$\dot{x} = f(x) + g(x)\kappa_i(x).$$

Furthermore, we have

$$J(x_0, \kappa_i(x(\cdot))) = V_i(x_0), \quad \forall x_0 \in \mathbb{R}^n,$$

which implies $V^*(x) \le V_i(x)$ for all $x \in \mathbb{R}^n$ and all $i \ge 0$.

2. *V_i is monotonically decreasing, i.e. $V_{i+1}(x) \le V_i(x)$ for all $x \in \mathbb{R}^n$ and $i \ge 0$, and $V_i(x) \to V^*(x)$ and $\kappa_i(x) \to \kappa^*(x)$, as $i \to \infty$, for all $x \in \mathbb{R}^n$. The convergence is uniform on every compact set of $\mathbb{R}^n$.*

Proof:

1. We show this by induction. Suppose that $u = k_i(x)$ is globally stabilizing, i.e. closed-loop solutions under κ_i approach A as $t \to \infty$. Let $x(t)$ denote such a solution and V_i satisfy (3.8). We have

$$V_i(x(t)) - V(x_0) = \int_0^t \frac{dV_i}{dx}(x(s))(f(x(s)) + g(x(s))\kappa_i(x(s)))ds$$

$$= -\int_0^t [Q(x(s)) + \kappa_i^T(x(s))R(x(s))\kappa_i(x(s))]ds,$$

which, by letting $t \to \infty$, implies

$$V_i(x_0) = J(x_0, \kappa_i). \tag{3.11}$$

Since κ_i is globally stabilizing, by (3.10) and (3.11), we have

$$V^*(x_0) = \min_{u(\cdot) \in \mathcal{K}(x_0)} J(x_0, u(\cdot)) \le J(x_0, \kappa_i) = V_i(x_0), \quad \forall x_0 \in \mathbb{R}^n.$$

Since V^* is assumed to be radially unbounded and positive definite, so is V_i. By (3.8), we have

$$\frac{\partial V_i}{\partial x}(f(x) + g(x)\kappa_i(x)) = -Q(x) - \kappa_i^T(x)R(x)\kappa_i(x) \le -Q(x), \tag{3.12}$$

which is negative definite. Hence, by Theorem 1.1, A is GUAS for the closed-loop system under $u = \kappa_i(x)$.

To complete the induction, we need to verify that $u = \kappa_{i+1}(x)$ is also stabilizing. We do so by verifying that V_i is also a Lyapunov function for the closed-loop system under $u = \kappa_{i+1}(x)$ given by (3.9). In view of (3.12) and the fact that $u = \kappa_{i+1}(x)$ minimizes the following quadratic function in u:

$$\frac{\partial V_i}{\partial x}(f(x) + g(x)u) + Q(x) + u^T R(x)u,$$

we have

$$\frac{\partial V_i}{\partial x}(f(x) + g(x)\kappa_{i+1}(x))$$

$$\le \frac{\partial V_i}{\partial x}(f(x) + g(x)\kappa_i(x)) + \kappa_i^T(x)R(x)\kappa_i(x) - \kappa_{i+1}^T(x)R(x)\kappa_{i+1}(x)$$

$$\le -Q(x) - \kappa_i^T(x)R(x)\kappa_i(x) + \kappa_i^T(x)R(x)\kappa_i(x) - \kappa_{i+1}^T(x)R(x)\kappa_{i+1}(x)$$

$$\le -Q(x) - \kappa_{i+1}^T(x)R(x)\kappa_{i+1}(x) \tag{3.13}$$

$$\le -Q(x),$$

which is negative definite. The proof is complete by Theorem 1.1.

2. For each $i \ge 0$, by (3.12) and (3.13), we have

$$\frac{\partial V_i}{\partial x}(f(x) + g(x)\kappa_{i+1}(x)) \le \frac{\partial V_{i+1}}{\partial x}(f(x) + g(x)\kappa_{i+1}(x)),$$

which, by substituting a solution $x(t)$ of the closed-loop system under $u = \kappa_{i+1}(x)$ and integrating over $[0, t]$, implies

$$V_i(x(t)) - V_i(x_0) = \frac{\partial V_i}{\partial x}(x(s))(f(x(s)) + g(x(s))\kappa_{i+1}(x(s)))$$

$$\le \frac{\partial V_{i+1}}{\partial x}(x(s))(f(x(s)) + g(x(s))\kappa_{i+1}(x(s)))$$

$$= V_{i+1}(x(t)) - V_{i+1}(x_0).$$

Letting $t \to \infty$, we obtain $V_i(x_0) \ge V_{i+1}(x_0)$ for all $x_0 \in \mathbb{R}^n$. Since $\{V_i(x)\}$ is monotonically decreasing and bounded below by V^*, it converges pointwise to a limit function $\overline{V}$. By Assumption 3.1 and a well-known theorem on uniform convergence and differentiation of sequence of functions (cf. Theorem 7.17 Rudin [1976] for the single variable case), $\{V_i\}$ converges uniformly to $\overline{V}$ on any compact set of $\mathbb{R}^n$. Moreover, we have

$$\frac{\partial \overline{V}}{\partial x} = \frac{\partial}{\partial x} \lim_{i \to \infty} V_i = \lim_{i \to \infty} \frac{\partial V_i}{\partial x}.$$

It then follows from (3.8) and (3.9) that $\overline{V}$ satisfies the HJB equation (3.5). Furthermore, $\overline{V}$ is positive definite and radially unbounded because it is lower bounded by V^*. By Remark 3.2, we have $\overline{V} = V^*$ by a uniqueness argument. ∎

Remark 3.3 While Assumption 3.1(3) may seem unnecessary, we do need to show that the gradient of the limit of V_i is indeed the limit of the gradient of V_i. This is not guaranteed by the mere convergence (or uniform convergence) of V_i to a limit (cf. the analysis in Saridis and Lee [1979]). For example, $f_n(x) = \frac{\sin(nx)}{n}$ converges uniformly to $f = 0$. However, $f_n'(x) = \cos(nx)$ does not converge. One can even construct a monotonically decreasing sequence with this property, e.g. $f_n(x) = \frac{\sin(4^n x) + 2}{4^n}$. It may be possible to remove or weaken Assumption 3.1(3) by exploiting the fact that V_i is the value function for the policy k_i, i.e. Eq. (3.11), which can be written explicitly as follows:

$$V_i(x_0) = \int_0^\infty [Q(x(s)) + \kappa_i^T(x(s))R(x(s))\kappa_i(x(s))]ds.$$

Differentiating this with regards to x_0 under the integral using differentiability with respect to the initial condition, x_0 can potentially be used to show that $\left\{ \frac{\partial V_i}{\partial x} \right\}$ converges uniformly on every compact subset of $\mathbb{R}^n$ or converges to $\frac{\partial \overline{V}}{\partial x}$ directly, where $\overline{V}$ is the limit of $\{V_i\}$. Stronger regularity assumptions on Q and R may be required. Fortunately for linear systems, verifying Assumption 3.1 is straightforward as shown in Section 3.2.1

3.2.1 Linear Quadratic Regulator

Consider the Linear Time-Invariant (LTI) system

$$\dot{x} = Ax + Bu, \quad x(t_0) = x_0, \tag{3.14}$$

with the cost

$$J(x_0, u) = \int_0^\infty [x^T(t)Qx(t) + u^T(t)Ru(t)]dt, \tag{3.15}$$

where both $Q \in \mathbb{R}^{n \times n}$ and $R \in \mathbb{R}^{m \times m}$ are symmetric and positive definite matrices. As usual, we denote this by $Q > 0$ and $R > 0$.

The exact policy iteration algorithm reduces to the following special case. Starting with a stabilizing gain matrix $K_0 \in \mathbb{R}^{n \times m}$ such that $A + BK_0$ is Hurwitz, for each $i \geq 0$, the algorithm does the following:

1. (Policy evaluation) Compute a symmetric positive definite matrix P_i by solving

$$Q + K_i^T R K_i + P_i(A + BK_i) + (A + BK_i)^T P_i = 0. \tag{3.16}$$

2. (Policy improvement) Update the policy

$$K_{i+1} = -R^{-1}B^T P_i. \tag{3.17}$$

We note that (3.16) and (3.17) readily imply (3.8) and (3.9), respectively, with $V_i(x) = x^T P_i x$. Assumption 3.1 can be easily verified as follows: Item (1) follows from the assumption that K_0 is stabilizing. Item (2) reduces to statement that the Algebraic Riccati Equation (ARE) (2.37), recalled here as follows:

$$Q + PA + A^T P - PBR^{-1}B^T P = 0,$$

has a unique symmetric and positive definite solution. According to Remark 2.2, this ARE has a unique stabilizing solution $\overline{P}$ that is symmetric and positive semidefinite if and only if (A, B) is stabilizable and (A, Q) is detectable. Furthermore, Proposition 2.7, if (A, Q) is observable, then $\overline{P}$ is also positive definite. Since we assume $Q > 0$, (A, Q) is clearly observable. Item (3) follows from the fact (3.16) reduces to the celebrated *Lyapunov equation*, which has a unique positive definite solution P_i, because $A + BK_i$ is Hurwitz and $Q + K_i^T R K_i$ is positive definite. If we can show that $\{P_i\}$ converges, then Assumption 3.1(3) holds. The above algorithm is summarized in Algorithm 3.2 (exact policy iteration for LQR). Theorem 3.2 restricted to this setting should guarantee that $P_i \to \overline{P}$ and $K_i \to -R^{-1}B^T\overline{P}$ as $i \to \infty$.

Algorithm 3.2 Exact Policy Iteration for LQR

Require: A, B, K_0
 1: **repeat**
 2:   Compute P_i such that (3.16) holds
 3:   Update K_{i+1} according to (3.17)
 4:   $i = i + 1$
 5: **until** $P_i = P_{i-1}$

The above discussion is summarized into the following result:

Corollary 3.1 *Suppose that (A, B) is stabilizable, $Q > 0$, and $R > 0$. Assume there exists K_0 such that $A + BK_0$ is Hurwitz. Let $\{P_i\}$ and $\{K_i\}$ be returned by Algorithm 3.2. We have*

$$\lim_{i \to \infty} P_i = \overline{P}, \quad \lim_{i \to \infty} K_i = -R^{-1}B^T\overline{P},$$

where $\overline{P}$ is the unique symmetric and positive definite solution of the ARE (2.37).

Proof: This is an immediate corollary of Theorem 3.2 by noting that $V_i(x) = x^T P_i x$ converges pointwise if and only if $\{P_i\}$ converges. Then Assumption 3.1 is verified and the conclusion follows from that of Theorem 3.2. ∎

3.3 Policy Iteration with Unknown Dynamics and Function Approximations

To apply policy iteration as in Algorithms 3.1 or 3.2, one needs to know the model, i.e. the pairs of functions (f, g) or matrices (A, B). Furthermore, for nonlinear systems, we may not be able to exactly represent the values $\{V_i\}$ and the controllers $\{\kappa_i\}$. In this section and section 3.3.1, we introduce approaches [Jiang and Jiang, 2012, 2014, 2017] that can overcome these drawbacks by computing approximate values and controls through measurement data.

To avoid direct use of the dynamic model (f, g), we rely on an initial stabilizing controller κ_0 for the system to generate measurement data that can be used to iteratively compute approximations of the value functions $\{V_i\}$ and controllers $\{\kappa_i\}$. We assume that κ_0 is locally Lipschitz and renders the closed-loop system

$$\dot{x} = f(x) + g(x)\kappa_0(x) \tag{3.18}$$

globally asymptotically stable in the sense that Assumption 3.1(1) is satisfied. For effectively learning the model from data, an artificial exploration noise ξ is often injected to the system through

$$\dot{x} = f(x) + g(x)(\kappa_0(x) + \xi). \tag{3.19}$$

To carry function approximations by state measurements, we shall assume that the class of exploration noise signals is bounded, and there exists a compact set $\Omega \subseteq \mathbb{R}^n$ that is forward invariant for system (3.19).

The main technical assumptions that enable policy iteration without exact models and exact function representations are summarized as follows:

Assumption 3.2 *In addition to Assumption 3.1, we assume the following:*

1. *There exists a compact set $\Omega \subseteq \mathbb{R}^n$ that is forward invariant for system (3.19) subject to the class of exploration noise signals used.*
2. *There exist infinite sequences of basis functions $\{\phi_j(x)\}$ and $\{\psi_j(x)\}$ such that each $V_i(x)$ and κ_{i+1} for $i \geq 0$ can be written as follows:*

$$V_i(x) = \sum_{j=1}^{\infty} v_{i,j}\phi_j(x), \quad \kappa_{i+1}(x) = \sum_{j=1}^{\infty} c_{i,j}\psi_j(x), \quad \forall x \in \Omega.$$

We assume that the convergence of these two series is uniform on Ω.

An approximate policy iteration algorithm can be obtained as follows: with the initial controller κ_0 and an exploration noise signal ξ, we can collect state measurements on intervals of the form $[t_k, t_k + \tau]$, $k = 1, 2, \ldots, N$. To avoid direct use of the model information in policy evaluation, we replace (3.8) with an integral form along the trajectory on $[t_k, t_k + \tau]$:

$$V_i(x(t_k + \tau)) - V_i(x(t_k))$$

$$= \int_{t_k}^{t_k+\tau} \frac{\partial V_i}{\partial x}(f(x) + g(x)(\kappa_0 + \xi))dt$$

$$= \int_{t_k}^{t_k+\tau} \frac{\partial V_i}{\partial x}(f(x) + g(x)\kappa_i)dt + \int_{t_k}^{t_k+\tau} \frac{\partial V_i}{\partial x}g(x)(\kappa_0 + \xi - u_i)dt$$

$$= \int_{t_k}^{t_k+\tau} [-Q(x) - \kappa_i^T(x)R(x)\kappa_i(x)]dt - 2\int_{t_k}^{t_k+\tau} \kappa_{i+1}^T(x)R(x)\mu_i dt, \qquad (3.20)$$

where $\mu_i = \kappa_0 + \xi - \kappa_i$. In (3.20), we used the fact that V_i satisfies (3.8) in the second last equality and $\frac{\partial V_i}{\partial x}g(x) = -2\kappa_{i+1}^T R(x)$ from (3.9). This allows us to remove the dependence on model information (f, g).

We would like to compute approximations of V_i and κ_{i+1} using (3.20). This is achieved by replacing V_i and κ_{i+1} with function approximations of the form

$$\hat{V}_i(x) = \sum_{j=1}^{p} \hat{v}_{i,j}\phi_j(x), \quad \hat{\kappa}_{i+1}(x) = \sum_{j=1}^{q} \hat{c}_{i,j}\psi_j(x), \qquad (3.21)$$

where the coefficients $\hat{v}_{i,j}$ and $\hat{c}_{i,j}$ are to be determined using (3.20). Let $\hat{\kappa}_i$ be given. With the definition of $\hat{V}_i$ and $\hat{\kappa}_{i+1}$, we can rewrite (3.20) as follows:

$$\sum_{j=1}^{p} \hat{v}_{i,j}[\phi_j(x(t_k + \tau)) - \phi_j(x(t_k))]$$

$$= \int_{t_k}^{t_k+\tau} [-Q(x) - \hat{\kappa}_i^T R(x)\hat{\kappa}_i]dt - \sum_{j=1}^{q} \hat{c}_{i,j} \int_{t_k}^{t_k+\tau} 2\psi_j^T R(x)\hat{\mu}_i dt, \qquad (3.22)$$

where $\hat{\mu}_i = \kappa_0 + \xi - \hat{\kappa}_i$. This is a linear equation with unknowns $\{\hat{v}_{i,j}\}$ and $\{\hat{c}_{i,j}\}$ to be determined. We have a total of N such equations due to the data collected over $[t_k, t_k + \tau]$, $k = 1, 2, \ldots, N$. Collectively, they can be recast in matrix form as follows:

$$\hat{W}_i \Phi_i = \Theta_i, \qquad (3.23)$$

where

$$\hat{W}_i = \begin{bmatrix} \hat{v}_{i,1} & \cdots & \hat{v}_{i,p} & \hat{c}_{i,1} & \cdots & \hat{c}_{i,q} \end{bmatrix},$$

$$\Phi_i = \begin{bmatrix} \phi_1(x(t_1 + \tau)) - \phi_1(x(t_1)) & \cdots & \phi_1(x(t_N + \tau)) - \phi_1(x(t_N)) \\ \vdots & \vdots & \vdots \\ \phi_p(x(t_1 + \tau)) - \phi_p(x(t_1)) & \cdots & \phi_p(x(t_N + \tau)) - \phi_p(x(t_N)) \\ \int_{t_1}^{t_1+\tau} 2\psi_1^T R(x)\hat{\mu}_i dt & \cdots & \int_{t_N}^{t_N+\tau} 2\psi_1^T R(x)\hat{\mu}_i dt \\ \vdots & \vdots & \vdots \\ \int_{t_1}^{t_1+\tau} 2\psi_q^T R(x)\hat{\mu}_i dt & \cdots & \int_{t_N}^{t_N+\tau} 2\psi_q^T R(x)\hat{\mu}_i dt \end{bmatrix},$$

and

$$\Theta_i = \left[\int_{t_1}^{t_1+\tau} [-Q(x) - \hat{\kappa}_i^T R(x)\hat{\kappa}_i]dt \cdots \int_{t_k}^{t_k+\tau} [-Q(x) - \hat{\kappa}_i^T R(x)\hat{\kappa}_i]dt \right].$$

Note the dimensions of $W_i \in \mathbb{R}^{1\times(p+q)}$, $\Phi \in \mathbb{R}^{(p+q)\times N}$, and $\Theta \in \mathbb{R}^{1\times N}$. The coefficients W_i can be determined as the least-square solution that minimizes the residual error:

$$\hat{W}_i = \underset{W\in\mathbb{R}^{1\times(p+q)}}{\arg\min} \| W\Phi_i - \Theta_i \|^2. \tag{3.24}$$

While this may yield multiple solutions, any one suffices for our purpose of minimizing the Euclidean norm of the residual error. In general, one such minimizing solution is given by $\hat{W}_i = \Theta_i \Phi_i^+$, where Φ_i^+ is the *Moore–Penrose pseudoinverse*. If Φ_i has full row rank $p + q$, $\hat{W}_i$ can be uniquely determined by $\hat{W}_i = \Theta_i \Phi_i^T(\Phi_i\Phi_i^T)^{-1}$, where $\Phi_i^T(\Phi_i\Phi_i^T)^{-1}$ is a right inverse that coincides with the Moore–Penrose pseudoinverse Φ_i^+. The following assumption spells out this full row rank assumption in a slightly stronger form.

Assumption 3.3 *There exists some $\rho > 0$ such that, for any $p \geq 1$, $q \geq 1$, and $i \geq 0$, there exists an $N \geq 1$ such that*

$$\frac{1}{N}\Phi_i\Phi_i^T - \rho I_{p+q} \geq 0.$$

Theorem 3.3 *Suppose that Assumptions 3.1, 3.2, and 3.3 hold. Given κ_0, let $\{V_i\}_{i=0}^{\infty}$ and $\{\kappa_{i+1}\}_{i=0}^{\infty}$ be obtained using (3.8) and (3.9). For any $i \geq 0$ and $\varepsilon > 0$, we can choose $p \geq 1$ and $q \geq 1$ sufficiently large such that*

$$\left| \hat{V}_i(x) - V_i(x) \right| \leq \varepsilon, \quad \left| \hat{\kappa}_{i+1}(x) - \kappa_{i+1}(x) \right| \leq \varepsilon, \quad \forall x \in \Omega,$$

where $\{\hat{V}_i\}$ and $\{\hat{\kappa}_{i+1}\}_{i=0}^{\infty}$ are updated using (3.24) with $\hat{\kappa}_0 = \kappa_0$.

Proof: We prove it by induction. For $i = 0$, we have $\hat{\kappa}_0 = \kappa_0$. Let

$$V_0(x) = \sum_{j=1}^{\infty} v_{0j}\phi_j(x), \quad \kappa_1(x) = \sum_{j=1}^{\infty} c_{0j}\psi_j(x), \quad \forall x \in \Omega.$$

For each $k \in \{1, \ldots, N\}$, by (3.20), we have

$$\sum_{j=1}^{\infty} v_{0j}[\phi_j(x(t_k + \tau)) - \phi_j(x(t_k))]$$

$$= \int_{t_k}^{t_k+\tau} [-Q(x) - \hat{\kappa}_0^T R(x)\hat{\kappa}_0]dt - \sum_{j=1}^{\infty} c_{0j} \int_{t_k}^{t_k+\tau} 2\psi_j^T R(x)\hat{\mu}_0 dt, \tag{3.25}$$

where $\hat{\mu}_0 = \kappa_0 - \hat{\kappa}_0 + \xi = \xi$, which can be rewritten as follows:

$$\sum_{j=1}^{p} v_{0j}[\phi_j(x(t_k + \tau)) - \phi_j(x(t_k))]$$

$$+ \sum_{j=1}^{q} c_{0j} \int_{t_k}^{t_k+\tau} 2\psi_j^T R(x)\hat{\mu}_0 dt + \int_{t_k}^{t_k+\tau} [Q(x) + \hat{\kappa}_0^T R(x)\hat{\kappa}_0] dt$$

$$= \sum_{j=p+1}^{\infty} v_{0j}[\phi_j(x(t_k + \tau)) - \phi_j(x(t_k))]$$

$$- \sum_{j=q+1}^{\infty} c_{0j} \int_{t_k}^{t_k+\tau} 2\psi_j^T R(x)\hat{\mu}_i dt := \delta_{0,k}. \qquad (3.26)$$

In matrix form, we can write

$$W_0 \Phi_0 - \Theta_0 = \Delta_0,$$

where

$$W_0 = \begin{bmatrix} v_{0,1} & \cdots & v_{0,p} & c_{0,1} & \cdots & c_{0,q} \end{bmatrix}, \quad \Delta_0 = \begin{bmatrix} \delta_{0,1} & \cdots & \delta_{0,N} \end{bmatrix}.$$

Write

$$\hat{W}_0 \Phi_0 - \Theta_0 = \Xi_0.$$

Since $\hat{W}_0$ is the least-square solution given by (3.24), we have $\|\Xi_0\|^2 \leq \|\Delta_0\|^2$. It follows that

$$\left\| (W_0 - \hat{W}_0)\Phi_0 \right\|^2 = (W_0 - \hat{W}_0)\Phi_0 \Phi_0^T (W_0 - \hat{W}_0)^T = \|\Delta_0 - \Xi_0\|^2 \leq 4\|\Delta_0\|^2,$$

which, by Assumption 3.3, implies

$$\left\| W_0 - \hat{W}_0 \right\|^2 \leq \frac{4\|\Delta_0\|^2}{N\rho} \leq \frac{4 \max_{1 \leq k \leq N} \delta_{0,k}^2}{\rho}.$$

We have $\lim_{p,q \to \infty} \delta_{0,k} = 0$ uniformly in k, which follows from

$$\sum_{j=p+1}^{\infty} v_{0j}\phi_j(x) \to 0, \quad \sum_{j=q+1}^{\infty} c_{1j}\psi_j(x) \to 0$$

uniformly on Ω as $p, q \to \infty$. From this and the fact that $\{\phi_j(x)\}$ is uniformly bounded on Ω, it follows that

$$\left| V_0(x) - \hat{V}_0(x) \right| = \left| \sum_{j=1}^{p} (v_{0j} - \hat{v}_{0j})\phi_j(x) + \sum_{j=p+1}^{\infty} v_{0j}\phi_j(x) \right|$$

$$\leq \sum_{j=1}^{p} \left| v_{0j} - \hat{v}_{0j} \right| \left| \phi_j(x) \right| + \left| \sum_{j=p+1}^{\infty} v_{0j}\phi_j(x) \right|$$

$$\leq \frac{\varepsilon}{2} + \frac{\varepsilon}{2} = \varepsilon, \quad \forall x \in \Omega,$$

provided that p and q are chosen sufficiently large. Similarly, we obtain $|\kappa_1(x) - \hat{\kappa}_1(x)| \leq \varepsilon$.

Now, suppose that the conclusion holds for some $i - 1 \geq 0$. Similar to (3.25) and (3.26), we obtain

$$\sum_{j=1}^{\infty} v_{i,j}[\phi_j(x(t_k + \tau)) - \phi_j(x(t_k))]$$

$$= \int_{t_k}^{t_k+\tau} [-Q(x) - \hat{\kappa}_i^T R(x)\hat{\kappa}_i]dt - \sum_{j=1}^{\infty} c_{i+1,j} \int_{t_k}^{t_k+\tau} 2\psi_j^T R(x)\hat{\mu}_i dt$$

$$+ \int_{t_k}^{t_k+\tau} \frac{\partial V_i}{\partial x} g(x)(\hat{\kappa}_i - \kappa_i)dt + \int_{t_k}^{t_k+\tau} [\hat{\kappa}_i^T R(x)\hat{\kappa}_i - \kappa_i^T R(x)\kappa_i]dt, \quad (3.27)$$

where $\hat{\mu}_i = \kappa_0 - \hat{\kappa}_i + \xi$, and

$$\sum_{j=1}^{p} v_{i,j}[\phi_j(x(t_k + \tau)) - \phi_j(x(t_k))]$$

$$+ \sum_{j=1}^{q} c_{i+1,j} \int_{t_k}^{t_k+\tau} 2\psi_j^T R(x)\hat{\mu}_0 dt + \int_{t_k}^{t_k+\tau} [Q(x) + \hat{\kappa}_i^T R(x)\hat{\kappa}_i]dt$$

$$= \delta_{i,k} + \gamma_{i,k}. \quad (3.28)$$

with

$$\delta_{i,k} = \sum_{j=p+1}^{\infty} v_{i,j}[\phi_j(x(t_k + \tau)) - \phi_j(x(t_k))] - \sum_{j=q+1}^{\infty} c_{i+1,j} \int_{t_k}^{t_k+\tau} 2\psi_j^T R(x)\hat{\mu}_i dt,$$

and

$$\gamma_{i,k} = \int_{t_k}^{t_k+\tau} \frac{\partial V_i}{\partial x} g(x)(\hat{\kappa}_i - \kappa_i)dt + \int_{t_k}^{t_k+\tau} [\hat{\kappa}_i^T R(x)\hat{\kappa}_i - \kappa_i^T R(x)\kappa_i]dt.$$

Using the same argument as in the base case, we have

$$\left\| W_i - \hat{W}_i \right\|^2 \leq \frac{4\|\Delta_i + \Gamma_i\|^2}{N\rho} \leq \frac{8(\max_{1 \leq k \leq N} \delta_{i,k}^2 + \max_{1 \leq k \leq N} \gamma_{i,k}^2)}{\rho},$$

where

$$W_i = \begin{bmatrix} v_{i,1} & \cdots & v_{i,p} & c_{i,1} & \cdots & c_{i,q} \end{bmatrix}, \quad \Delta_i = \begin{bmatrix} \delta_{i,1} & \cdots & \delta_{i,N} \end{bmatrix},$$

$$\Gamma_i = \begin{bmatrix} \gamma_{i,1} & \cdots & \gamma_{i,N} \end{bmatrix}.$$

We have $\lim_{p,q \to \infty} \delta_{i,k} = 0$ uniformly in k as before and $\lim_{p,q \to \infty} \gamma_{i,k} = 0$ uniformly in k follows from the inductive assumption. We can now show that the conclusion holds for i in the same way. ∎

Note that Assumption 3.2(1) is essential in ensuring that all trajectory data are collected within the compact set Ω so that all functions involved are bounded for some of the limit argument in the above proof to go through.

Intuitively, the above theorem states that, if we choose a sufficiently large number of basis functions and obtain sufficiently rich measurement data such that Assumption 3.3 holds, then we can approximate, to an arbitrary degree of accuracy, the value functions $\{V_i\}$ and controllers $\{\kappa_i\}$ obtained from the exact policy iteration Algorithm 3.1.

3.3.1 Linear Quadratic Regulator with Unknown Dynamics

In this section, we discuss the special case of approximate policy iteration for linear systems with unknown dynamics [Jiang and Jiang, 2012]. We recall the LTI system

$$\dot{x} = Ax + Bu, \quad x(t_0) = x_0, \tag{3.29}$$

with the cost

$$J(x_0, u) = \int_0^\infty [x^T(t)Qx(t) + u^T(t)Ru(t)]dt, \tag{3.30}$$

where $Q > 0$ and $R > 0$. Let K_0 be given such that $A + BK_0$ is Hurwitz.

As discussed in Section 3.2.1, in this case $V_i(x) = x^T P_i x$ and $\kappa(x) = K_i x$, where $P_i \in \mathbb{R}^{n \times n}$ is a symmetric positive definite matrix and $K_i \in \mathbb{R}^{n \times m}$. Policy iteration reduces to computation of P_i by successively solving the Lyapunov equation (3.16) and updating K_i according to (3.17).

The quadratic function $V_i(x) = x^T P_i x$ can be written as a finite linear combination of basis functions using a vectorization trick:

$$V_i(x) = x^T P_i x = \text{vec } (P_i)^T(x \otimes x), \tag{3.31}$$

where vec$(\cdot)$ is the *vectorization* of a matrix to a column vector, by stacking the columns of the matrix in order, and $\otimes$ is the *Kronecker product*. Clearly, vec $(P_i)^T$ is a row vector and $x \otimes x$ is a column vector that contains all quadratic and linear monomials of coefficient one, formed by components of x. Similarly, we can write

$$x^T K_i^T = \text{vec } (K_i)^T(x \otimes I_m). \tag{3.32}$$

With (3.31) and (3.32), (3.22) can be written as follows:

$$\text{vec } (P_i)^T(x(t_k + \tau) \otimes x(t_k + \tau) - x(t_k) \otimes x(t_k))$$
$$= \int_{t_k}^{t_k+\tau} x^T[-Q - K_i^T RK_i]x dt - \text{vec } (K_{i+1})^T \int_{t_k}^{t_k+\tau} 2(x \otimes I_m)R\mu_i dt, \tag{3.33}$$

where $\mu_i = (K_0 x - K_i x + \xi)$. Note that, since P_i and K_{i+1} have a finite number of unknowns, we aim to solve them exactly instead of approximating them as in the nonlinear case. Similar to (3.23), we write all N equations for the unknowns in vec(P_i) and vec(K_{i+1}) in matrix form as follows:

$$\left[\text{vec}(P_i) \ \text{vec}(K_{i+1})\right] \Phi_i = \Theta_i, \tag{3.34}$$

where

$$\Phi_i = \begin{bmatrix} x(t) \otimes x(t)|_{t_1}^{t_1+\tau} & \cdots & x(t) \otimes x(t)|_{t_N}^{t_N+\tau} \\ \int_{t_1}^{t_1+\tau} 2(x \otimes I_m)R\mu_i dt & \cdots & \int_{t_N}^{t_N+\tau} 2(x \otimes I_m)R\mu_i dt \end{bmatrix},$$

and

$$\Theta_i = \begin{bmatrix} \int_{t_1}^{t_1+\tau} x^T[-Q - K_i^T RK_i]x dt & \cdots & \int_{t_k}^{t_k+\tau} x^T[-Q - K_i^T RK_i]x dt \end{bmatrix}.$$

Note that the dimension of Φ_i is $(n^2 + mn) \times N$. Clearly, if $\text{rank}(\Phi) = n^2 + mn$, then $\text{vec}(P_i)$ and $\text{vec}(K_{i+1})$, which together have $n^2 + mn$ unknowns, can be uniquely determined. A closer look at the equations reveals that due to the symmetry of P_i, there are only $\frac{n(n+1)}{2} + mn$ unknowns. The positions of the duplicated unknowns in $\text{vec}(P_i)$ exactly correspond to the duplicated rows in Φ_i. As a result, the meaningful rank condition is $\text{rank}(\Phi_i) = \frac{n(n+1)}{2} + mn$. Once this is satisfied, we know (3.34) has a unique set of solutions (P_i, K_{i+1}). Inconsistency cannot occur, as we know that P_i can be uniquely solved from the Lyapunov equation (3.16) and K_{i+1} can be determined from (3.17).

This is summarized in the following result, which follows immediately from Corollary 3.1.

Corollary 3.2 *Let K_0 be such that $A + BK_0$ is Hurwitz. Suppose that for each $i \geq 0$, Φ_i in (3.34) satisfies $\text{rank}(\Phi_i) = \frac{n(n+1)}{2} + mn$. Then (P_i, K_{i+1}) is uniquely determined for each $i \geq 0$. Furthermore,*

$$\lim_{i\to\infty} P_i = \overline{P}, \quad \lim_{i\to\infty} K_i = -R^{-1}B^T\overline{P},$$

where $\overline{P}$ is the unique symmetric and positive definite solution of the ARE (2.37).

3.4 Summary

Policy iteration is a popular RL technique. In this chapter, we discussed the policy iteration method for continuous control problems. We showed that for systems that are affine in control and have a quadratic-in-control cost, the HJB equation can be used to devise policy iteration algorithms that iteratively improve the performance of a stabilizing controller with respect to the cost functional. In the nonlinear case, the nonlinear HJB equation is reduced to a linear Partial Differential Equation (PDE) at each iteration of policy evaluation. This result dates back to Milshtein [1964], Vaisbord [1963], Leake and Liu [1967], and Saridis and Lee [1979]. A computational framework was developed in Beard [1995]. In the linear case, the HJB equation is reduced to an ARE. The solution of the ARE can be approximated by iteratively solving Lyapunov equations that successively improve the quadratic value function. This algorithm is known as the Kleinman algorithm [Kleinman, 1968]. Section 3.3 was based on more recent work by Jiang and Jiang [2012], Jiang and Jiang [2014], and Jiang and Jiang [2017], where the dependence on model information in the earlier algorithms [Milshtein, 1964; Vaisbord, 1963; Leake and Liu, 1967; Kleinman, 1968; Saridis and Lee, 1979] was circumvented by seeking to determine the pair (V_i, κ_{i+1}) or (P_i, K_{i+1}) simultaneously from measurement data. This extends earlier work by Vrabie and Lewis [2009] and Bhasin et al. [2013] that partially requires model information (more specifically, g and B). We do not attempt to provide a comprehensive overview of RL for continuous control problems. The readers are referred to Jiang and Jiang [2017], Kamalapurkar et al. [2018], Vamvoudakis et al. [2021], Lewis et al. [2012], Busoniu et al. [2017], and Lewis and Liu [2013] and references therein.

Bibliography

Randal Winston Beard. *Improving the Closed-Loop Performance of Nonlinear Systems.* PhD thesis, Rensselaer Polytechnic Institute, 1995.

Richard E. Bellman. *Dynamic Programming.* Princeton University Press, 1957.

Dimitri P. Bertsekas. *Dynamic Programming and Optimal Control: Volume I*, volume 1. Athena Scientific, 2012.

Dimitri P. Bertsekas. *Reinforcement Learning and Optimal Control.* Athena Scientific, 2019.

Shubhendu Bhasin, Rushikesh Kamalapurkar, Marcus Johnson, Kyriakos G. Vamvoudakis, Frank L. Lewis, and Warren E. Dixon. A novel actor–critic–identifier architecture for approximate optimal control of uncertain nonlinear systems. *Automatica*, 49(1):82–92, 2013.

Lucian Busoniu, Robert Babuska, Bart De Schutter, and Damien Ernst. *Reinforcement Learning and Dynamic Programming Using Function Approximators.* CRC Press, 2017.

Ronald A. Howard. *Dynamic Programming and Markov Process.* MIT Press, 1960.

Yu Jiang and Zhong-Ping Jiang. Computational adaptive optimal control for continuous-time linear systems with completely unknown dynamics. *Automatica*, 48(10):2699–2704, 2012.

Yu Jiang and Zhong-Ping Jiang. Robust adaptive dynamic programming and feedback stabilization of nonlinear systems. *IEEE Transactions on Neural Networks and Learning Systems*, 25(5):882–893, 2014.

Yu Jiang and Zhong-Ping Jiang. *Robust Adaptive Dynamic Programming.* John Wiley & Sons, 2017.

Rushikesh Kamalapurkar, Patrick Walters, Joel Rosenfeld, and Warren Dixon. *Reinforcement Learning for Optimal Feedback Control.* Springer, 2018.

David Kleinman. On an iterative technique for Riccati equation computations. *IEEE Transactions on Automatic Control*, 13(1):114–115, 1968.

R. J. Leake and Ruey-Wen Liu. Construction of suboptimal control sequences. *SIAM Journal on Control*, 5(1):54–63, 1967.

Frank L. Lewis and Derong Liu. *Reinforcement Learning and Approximate Dynamic Programming for Feedback Control*. John Wiley & Sons, 2013.

Frank L. Lewis, Draguna Vrabie, and Vassilis L Syrmos. *Optimal Control*. John Wiley & Sons, 2012.

G. N. Milshtein. Successive approximations for solution of one optimal problem. *Automation and Remote Control*, 25(3):298–306, 1964.

Walter Rudin. *Principles of Mathematical Analysis*. McGraw-Hill, 3rd edition, 1976.

George N. Saridis and Chun-Sing G. Lee. An approximation theory of optimal control for trainable manipulators. *IEEE Transactions on Systems, Man, and Cybernetics*, 9(3):152–159, 1979.

Richard S. Sutton and Andrew G. Barto. *Reinforcement Learning: An Introduction*. MIT Press, 2018.

E. M. Vaisbord. Concerning an approximate method for optimum control synthesis. *Avtomatika i Telemekhanika*, 24(12):1626–1632, 1963.

Kyriakos G. Vamvoudakis, Yan Wan, Frank L. Lewis, and Derya Cansever. *Handbook of Reinforcement Learning and Control*. Springer, 2021.

Draguna Vrabie and Frank Lewis. Neural network approach to continuous-time direct adaptive optimal control for partially unknown nonlinear systems. *Neural Networks*, 22(3):237–246, 2009.

4

Learning of Dynamic Models

The Model-based Reinforcement Learning (MBRL) techniques we shall discuss in later chapters rely on identifying a dynamic model of the system. *System identification* can be defined as a set of techniques used for obtaining a model of a dynamical system based on the observations and our prior knowledge of the system. In practice, the model acquired almost always represents only an approximation of the real plant. We give a brief account of such approximation techniques in this chapter.

4.1 Introduction

4.1.1 Autonomous Systems

Consider the autonomous system given by

$$\dot{x} = f(x), \tag{4.1}$$

where $x \in D \subseteq \mathbb{R}^n$ and $f : D \to \mathbb{R}^n$. Assume that a sequence of state measurements (or sampled states) $\{x_k\}$, taken at times $\{t_k\}$, are available, where $k = 0, 1, \ldots, N_s$ and N_s denotes the total number of samples.

4.1.2 Control Systems

While in this chapter, we mainly present the learning algorithms for the autonomous system (4.1), similar techniques can be employed to identify a control system, with some considerations which will be discussed in more detail in Section 4.3.3. To this end, we consider the general control-affine system

$$\dot{x} = f(x) + g(x)u, \tag{4.2}$$

Model-Based Reinforcement Learning: From Data to Continuous Actions with a Python-based Toolbox, First Edition. Milad Farsi and Jun Liu.

where $u \in U \subseteq \mathbb{R}^m$, $f : D \to \mathbb{R}^n$, and $g : D \to \mathbb{R}^{n \times m}$. We assume that sampled controls $\{u_k\}$, together with the state measurements $\{x_k\}$, taken at different times $\{t_k\}$ are available.

The objective of this chapter is to learn a model that can best fit the training data and, more importantly, be used to efficiently predict the system's behaviors for unseen data points. Two elements are important for achieving this objective: data collection and model selection. Although the role of these two elements and their relation to each other may seem trivial, identifying the properties required for an efficient learning of a real-world process is a challenging task. Various constraints in applications might complicate the selection of an appropriate model. This task becomes even more challenging in the Reinforcement Learning (RL) setting, where additionally the data need to be obtained since it is not readily provided in the first place.

In the Section 4.2, we highlight the main concerns in the selection of the model and the data collection. Moreover, we discuss and compare the main techniques existing in the literature.

4.2 Model Selection

There exist different options in choosing a model to fit data, depending on the availability of data and computational resources. In this section, we categorize different models and discuss their advantages and drawbacks.

4.2.1 Gray-Box vs. Black-Box

In white-box methods, physical laws are used to directly derive the set of equations defining the system. Identification techniques can be roughly categorized into two different groups. On the one hand, in the gray-box identification, only the structure of the model is known, while there exist some unknown or uncertain parameters that need to be estimated through observations. Accordingly, the structure assumed may be a result of a white-box identification with uncertain parameters. On the other hand, the black-box technique utilizes a generic parameterization that is chosen independently from the physical laws and relationships underlying the system [Keesman, 2011].

4.2.2 Parametric vs. Nonparametric

Various supervised approaches are employed for learning of dynamical systems. In general, they can be categorized into two groups of techniques that have fundamental differences.

In the parametric approach, the parameters are taken to be of the highest importance for establishing the model. Hence, one needs to first decide how the model is parameterized. Then, come up with a technique for updating the parameters to best describe the data. Such techniques usually include a limited number of parameters in the model. Hence, it can be considered as a light-weight candidate for learning in real time. For this reason, they are the first choice in the majority of the adaptive control techniques. Linear and nonlinear regression techniques are some examples of parametric learning type.

Another factor that makes the parametric models more suitable is their explainability. This is a result of the brevity in the model that is in contrast with the considerably large size of nonparametric models that cannot be read and explained directly by human. This advantage also provides an opportunity for development of analytical frameworks based on the learned model.

On the other hand, nonparametric models directly rely on data for learning the unknown process. It is either that any single sample is used as a basis for the model or the number of parameters employed is extremely large that they cannot be tracked. A good example is the Gaussian processes that use the data to obtain a Bayesian inference. The probabilistic feature of such models helps in more efficient predictions, where one can also draw information on how confident the predictions are. Another strength of such methods is that they allow incorporation of the user knowledge by choosing a variety of kernels in different situations. However, almost all of the nonparametric techniques experience difficulty with the increasing size of the data. Hence, such approaches cannot scale well if a large number of data points need to be considered.

In fact, if a spectrum of the models can be assumed, small-sized parametric models and large-sized nonparametric models constitute the extreme cases. In this range, different Neural Networks (NNs) mostly fall in the middle, depending on their sizes. In general, NNs often include many weights and biases, which play the main role in determining the complexity of the computations. However, considering the fact that their complexity does not increase by the number of samples used, one can still see them as parametric models. On the other hand, the deeper and wider they grow, the number of the weights and biases tend to be excessively large, until there is no more advantages in looking at the model parameterwise. In this case, one may choose to treat them as nonparametric models.

Figure 4.1 demonstrates the main categories of models with some examples. Moreover, we draw a comparison of the complexity of these techniques in general, where for a fair comparison, one needs to also take into account the size of the data.

In what follows, we review some well-known techniques for updating a parameterized model through observations.

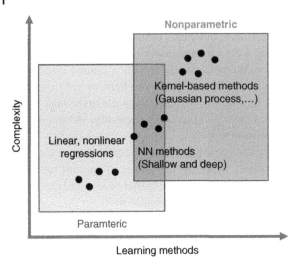

Figure 4.1 A view of parametric and nonparametric learning methods in terms of complexity.

4.3 Parametric Model

Consider the autonomous system (4.1). Given the measurements x_k, $\tilde{\dot{x}}_k$ can be obtained as an approximation of $\dot{x}_k$ by a numerical differentiation technique along the sequence. For this purpose, different methods exist in the literature for obtaining an accurate derivative using equally or unequally spaced sequences of data, see, e.g. Li [2005].

Having N_s number of samples of x_k and the corresponding $\tilde{\dot{x}}_k$ at our disposal, we can now choose a parametric model to fit the data collected:

$$\dot{x} = \hat{f}_w(x) + e,\tag{4.3}$$

where w represents the set of parameters to be trained, $\hat{f}_w(x)$ is an approximation of $f(x)$, and e is the prediction error.

4.3.1 Model in Terms of Bases

A popular configuration of the model is obtained by a linear combination of some differentiable basis functions. This is because of the linear dependency on the parameters that offers flexibility in training the parameters. Moreover, the obtained model can be easily handled for further system design and analysis:

$$\dot{x} = w\Phi(x) + e,\tag{4.4}$$

where $w \in \mathbb{R}^{n \times p}$ is the matrix of the weights, $\Phi : D \to \mathbb{R}^p$ is the vector of basis functions, and e is the prediction error.

4.3.2 Data Collection

One of the properties that contrasts the type of learning techniques discussed in this chapter with the usual machine learning techniques is the dynamic data. Therefore, if the learning algorithm requires a batch of input–output samples, the most one can do is to gradually grow a dataset, as samples become available over time.

For this purpose, we save the data consisting of sampled derivatives and bases functions evaluated at the state measurements as follows:

$$\dot{X} = \begin{bmatrix} \dot{x}_1 & \dot{x}_2 & \dots & \dot{x}_{N_s} \end{bmatrix}_{n \times N_s},$$

$$\Psi = \begin{bmatrix} \Phi(x_1) & \Phi(x_2) & \dots & \Phi(x_{N_s}) \end{bmatrix}_{p \times N_s}, \tag{4.5}$$

where each column corresponds to a single sample, and N_s denotes the total number of samples.

4.3.3 Learning of Control Systems

For the purpose of RL, we need to learn a control system rather than the autonomous system (4.1). Thus, the parametric model (4.4) and algorithms developed based on it cannot be directly used.

Therefore, we need to bring the measurements of the control in the model update procedure. In parametric model update, this can be done by including bases in terms of the control input. Assuming the control-affine system (4.2), one can choose the vector of bases as

$$\Phi^u(x, u) = \begin{bmatrix} \Phi_0(x) \\ \Phi_1(x)u_1 \\ \vdots \\ \Phi_m(x)u_m \end{bmatrix},$$

where $u_j \in \mathbb{R}$ and $\Phi_j(x)$ are the control inputs and some vector of bases in terms of x chosen, respectively, for $j = 1, \dots, m$. Then, assuming that the control input can also be sampled at any time in the same way as the state, one can now store the data in the following order:

$$\dot{X}^u = \begin{bmatrix} \dot{x}_1 & \dot{x}_2 & \dots & \dot{x}_{N_s} \end{bmatrix}_{n \times N_s},$$

$$\Psi^u = \begin{bmatrix} \Phi(x_1, u_1) & \Phi(x_2, u_2) & \dots & \Phi(x_{N_s}, u_{N_s}) \end{bmatrix}_{p \times N_s},$$

where each column corresponds to one sample.

By using such a modification in the vector of bases and accordingly in the datasets (4.5), we can implement the presented algorithms to obtain an estimation of the control system. However, in practice, an efficient estimation of the system also depends on the quality of data that will be discussed in Section 4.4.

4.4 Parametric Learning Algorithms

4.4.1 Least Squares

Let us assume the following loss by considering the square of errors:

$$J(w) = \frac{1}{2}\sum_{i=1}^{N_s} e_i^T e_i,$$

where e_i denotes the error resulting from each sample. In view of (4.4), J is a quadratic function of w. Let us consider the problem with respect to each jth component of the error. Then, to minimize the loss, we take the gradient with respect to the jth row of parameters, denoted by w_j, as follows:

$$\frac{\partial}{\partial w_j}J(w_j) = \frac{\partial}{\partial w_j}\left(\frac{1}{2}\sum_{i=1}^{N_s} e_{ij}^T e_{ij}\right)$$

$$= \frac{\partial}{\partial w_j}\left(\frac{1}{2}\sum_{i=1}^{N_s}(\dot{x}_{ij} - w_j\Phi(x_i))^T(\dot{x}_{ij} - w_j\Phi(x_i))\right)$$

$$= \sum_{i=1}^{N_s}(w_j\Phi(x_i) - \dot{x}_{ij})\Phi(x_i)^T$$

$$= w_j\sum_{i=1}^{N_s}\Phi(x_i)\Phi(x_i)^T - \sum_{i=1}^{N_s}\dot{x}_{ij}\Phi(x_i)^T$$

$$= 0,$$

where it should be noted that because of the linear dependency of e_{ij} on the parameters, the partial derivative with respect to the matrix of parameters is given in terms of the regressor vector $\Phi(x)$. Now, by solving the equation for the parameters, we can achieve an estimation of the jth row of the parameters corresponding to dynamic $\dot{x}_j$ in (4.4) as follows:

$$\hat{w}_j = \left(\sum_{i=1}^{N_s}\dot{x}_{ij}\Phi(x_i)^T\right)\left(\sum_{i=1}^{N_s}\Phi(x_i)\Phi(x_i)^T\right)^{-1}.$$

Now, if we consider all the rows $j = 1, \ldots, n$ together, we can write the estimation relation for the matrix of parameters w as follows:

$$\hat{w} = \left(\sum_{i=1}^{N_s}\dot{x}_i\Phi(x_i)^T\right)\left(\sum_{i=1}^{N_s}\Phi(x_i)\Phi(x_i)^T\right)^{-1}. \tag{4.6}$$

Accordingly, for any update in the parameters, one needs to process all the data collected in batch. Alternatively, all these computations can be done in the matrix form if data are prepared in the format (4.5). In this way, the weights can be simply obtained by matrix multiplications and pseudoinverse $\Psi^\dagger = \Psi^T \left(\Psi \Psi^T \right)^{-1}$ as

$$\hat{w} = \dot{X} \Psi^\dagger.$$

However, there is still no improvement made in the computational complexity since it is determined by the number of samples.

Algorithm 4.1 Least Squares Regression

1: Data: $\Psi, \dot{X}$
2: Output: $\hat{w}$
3: $\hat{w} \leftarrow \dot{X} \Psi^T \left(\Psi \Psi^T \right)^{-1}$

4.4.2 Recursive Least Squares

The least squares (LS) regression is an efficient technique to fit the model when the data are provided in batch. However, this is not an option in the RL setting, where the samples are collected one by one in an online scheme. Alternatively, one can possibly construct a growing list of samples starting from the first sample to the N_sth sample, where for each step, a batch LS is run to update the parameters. This technique may be applicable when N_s is not large. Otherwise, the computational complexity will grow fast by each observation, making the calculations infeasible in real time.

Another alternative is to obtain a recursive parameter update when each sample can be processed individually. In this way, after initialization of the parameters, each observation made can update the parameters independently. Hence, the complexity will remain the same for any step of learning. Later, in the simulation results, a comparison will demonstrate the difference of these algorithms.

To realize a recursive algorithm, let us start with the formula obtained for the batch mode that is given by (4.6). This time, we do not assume availability of a list of samples in the first place. Instead, we consider steps $k \in \{1, 2, \dots, N_s\}$ and obtain a stepwise variant of (4.6) as follows:

$$\hat{w}_k = \left(\sum_{i=1}^k \dot{x}_i \Phi(x_i)^T \right) \left(\sum_{i=1}^k \Phi(x_i) \Phi(x_i)^T \right)^{-1}. \tag{4.7}$$

Then, by defining R_k, we obtain a recursive relation for the second term in (4.7) as follows:

$$
\begin{aligned}
R_k &= \sum_{i=1}^k \Phi(x_i) \Phi(x_i)^T \\
&= \sum_{i=1}^{k-1} \Phi(x_i) \Phi(x_i)^T + \Phi(x_k) \Phi(x_k)^T \\
&= R_{k-1} + \Phi(x_k) \Phi(x_k)^T. \tag{4.8}
\end{aligned}
$$

Accordingly, we can rewrite the first term in (4.7) as

$$\sum_{i=1}^{k} \dot{x}_i \Phi(x_i)^T = \sum_{i=1}^{k-1} \dot{x}_i \Phi(x_i)^T + \dot{x}_k \Phi(x_k)^T$$

$$= \hat{w}_{k-1} R_{k-1} + \dot{x}_k \Phi(x_k)^T$$

$$= \hat{w}_{k-1} (R_k - \Phi(x_k)\Phi(x_k)^T) + \dot{x}_k \Phi(x_k)^T$$

$$= \hat{w}_{k-1} R_k + \left(\dot{x}_k - \hat{w}_{k-1} \Phi(x_k)\right) \Phi(x_k)^T.$$

Plugging in (4.7) yields a recursive relation for the parameters as follows:

$$\hat{w}_k = \hat{w}_{k-1} + \left(\dot{x}_k - \hat{w}_{k-1} \Phi(x_k)\right) \Phi(x_k)^T R_k^{-1}. \tag{4.9}$$

However, still the calculation of R_k requires all the samples. Considering that R_k can be written in terms of its previous value as in (4.8), the following lemma can be employed.

Lemma 4.1 *(Sherman–Morrison Formula [Sherman, 1949; Bartlett, 1951])*
Let $A \in \mathbb{R}^{n \times n}$ and $u, v \in \mathbb{R}^n$. Moreover, we assume the inverse A^{-1} exists such that $1 + v^T A^{-1} u \neq 0$. Denote $\bar{A} = A + uv^T$. Then the following holds:

$$\bar{A}^{-1} = A^{-1} - \frac{A^{-1} uv^T A^{-1}}{1 + v^T A^{-1} u}. \tag{4.10}$$

Proof: It is easy to verify that

$$\bar{A}\bar{A}^{-1} = (A + uv^T) \left(A^{-1} - \frac{A^{-1} uv^T A^{-1}}{1 + v^T A^{-1} u} \right) = I_n.$$

Hence, (4.10) is indeed the inverse of $\bar{A}$. However, a somewhat straightforward proof can be also given by

$$\bar{A}^{-1} = (A + uv^T)^{-1} = A^{-1}(1 + uv^T A^{-1})^{-1}$$

$$= A^{-1}(1 - uv^T A^{-1} + uv^T A^{-1} uv^T A^{-1} - \ldots)$$

$$= A^{-1} - A^{-1} uv^T A^{-1}(1 - v^T A^{-1} u + (v^T A^{-1} u)^2 - \ldots)$$

$$= A^{-1} - \frac{A^{-1} uv^T A^{-1}}{1 + v^T A^{-1} u}. \qquad \blacksquare$$

Now, using this lemma and (4.8), we obtain

$$R_k^{-1} = R_{k-1}^{-1} - \frac{R_{k-1}^{-1} \Phi(x_k)\Phi(x_k)^T R_{k-1}^{-1}}{1 + \Phi(x_k)^T R_{k-1}^{-1} \Phi(x_k)}. \tag{4.11}$$

Defining $P_k = R_k^{-1}$, this relation together with (4.9) constructs the Recursive Least Squares (RLS) algorithm.

A rather different result can be obtained if the loss is modified such that the error for the older data takes less importance compared to the recent ones. This can be tuned by a positive constant $\lambda < 1$, namely the forgetting factor, where values closer to one discourage forgetfulness and smaller values help forgetting older data [Ljung and Söderström, 1983]:

$$J(w) = \frac{1}{2} \sum_{i=1}^{k} \lambda^{k-i} e_i^T e_i.$$

In Algorithm 4.2, we summarize the RLS algorithm.

Algorithm 4.2 Recursive Least Squares

1: Data: $\Psi, \dot{X}$ $\triangleright$ Only column $\Phi(x_k)$ is accessed at each iteration k
2: Output: $\hat{w}$
3: Initialize:
4:     $P_0 \leftarrow \kappa I_{p \times p}$ $\triangleright \kappa \leftarrow$ a large positive value
5:     $w_0 \leftarrow 0_{n \times p}$
6:     $\lambda \leftarrow 0.99$ $\triangleright \lambda \in [0.95, 1)$ is recommended
7: **for** $k \leftarrow 1$ to N_s **do**
8:     $P_k \leftarrow \frac{1}{\lambda} \left(P_{k-1} - \frac{P_{k-1} \Phi(x_k) \Phi(x_k)^T P_{k-1}}{\lambda + \Phi(x_k)^T P_{k-1} \Phi(x_k)} \right)$
9:     $\hat{w}_k \leftarrow \hat{w}_{k-1} + \left(\dot{x}_k - \hat{w}_{k-1} \Phi(x_k) \right) \Phi(x_k)^T P_k$
10: **end for**

4.4.3 Gradient Descent

Gradient Descent (GD) is an iterative optimization algorithm that can be used to obtain an estimation of the model parameters. Assume a differentiable objective $J(w)$ given as follows:

$$J(w) = \frac{1}{2} e^T e. \tag{4.12}$$

Then, by moving in the direction of steepest decent as

$$\hat{w}_k = \hat{w}_{k-1} - \gamma \left(\partial J(w) / \partial w \right) \Big|_{\hat{w}_{k-1}, \Phi(x_k), \dot{x}_k}$$

$$= \hat{w}_{k-1} - \gamma e \frac{\partial e}{\partial w}^T \Big|_{\hat{w}_{k-1}, \Phi(x_k), \dot{x}_k}$$

$$= \hat{w}_{k-1} - \gamma (\hat{w}_{k-1} \Phi(x_k) - \dot{x}_k) \Phi(x_k)^T$$

for a small enough $\gamma \in \mathbb{R}^+$, we have $J(w_{k-1}) \leq J(w_k)$. Therefore, we hope the resulting monotonic sequence leads to a desired minimum. Considering that a single point is used to estimate the gradient, this approach often exhibits poor convergence results with random fluctuation. The similar idea is also used in adaptive filtering that is well known as Least Mean Squares (LMS) technique. The procedure is given in Algorithm 4.3.

Algorithm 4.3 Gradient Descent

1: Data: $\Psi, \dot{X}$ ▷ Only column $\Phi(x_k)$ is accessed at each iteration k
2: Output: $\hat{w}$
3: Initialize:
4:    $w_0 \leftarrow 0_{n \times p}$
5:    γ is set to be small enough
6: **for** $k \leftarrow 1$ to N_s **do**
7:    $\hat{w}_k \leftarrow \hat{w}_{k-1} - \gamma(\hat{w}_{k-1}\Phi(x_k) - \dot{x}_k)\Phi(x_k)^T$
8: **end for**

4.4.4 Sparse Regression

In the sparse identification techniques, in addition to minimizing the prediction error, we also minimize the magnitude of the parameters according to the loss defined as follows:

$$J(w) = \sum_{i=1}^{N_s} e_i^T e_i + \lambda \|w\|_1, \tag{4.13}$$

where $\lambda > 0$ is a weighting factor. This choice of the loss encourages the sparsity in the model. To solve such an optimization problem, we employ an iterative LSs technique that uses a threshold set by user to eliminate small values appeared in $\hat{w}$ as proposed in Brunton et al. [2016]. The procedure is provided in Algorithm 4.4.

4.5 Persistence of Excitation

Let us again consider the loss (4.12). This time, we assume that the process can be continuously measured. Hence, a coupled differential equation can be obtained to update $\hat{w}$. Using a continuous GD update rule, we derive

$$\dot{w} = -\gamma \frac{\partial}{\partial w} J(w) = -\gamma e \Phi(x)^T.$$

Let the optimal parameters be denoted by w^*. Then, given $w_e = w^* - \hat{w}$, as the estimation error, we have

$$\dot{w}_e = \dot{w} = -\gamma(\dot{x} - \hat{w}\Phi(x))\Phi(x)^T$$
$$= -\gamma(w^*\Phi(x) - \hat{w}\Phi(x))\Phi(x)^T$$
$$= -\gamma w_e \Phi(x)\Phi^T(x).$$

Algorithm 4.4 Sparse Regression

1: Data: $\Psi, \dot{X}, \lambda$ ▷ λ: a threshold set by user
2: Output: $\hat{w}$
3: Initialize:
4:    $\hat{w} \leftarrow \dot{X}\Psi^\dagger$ ▷ Obtain initial parameters using LS regression
5:    $k \leftarrow 0$
6: **while** not converged **do**
7:    $k + +$
8:    $I_{small} \leftarrow \text{where}(|\hat{w}| < \lambda)$ ▷ Find the indices with small components $\hat{w}$
9:    $\hat{w}[I_{small}] \leftarrow 0$ ▷ Set the corresponding parameters to zeros
10:    **for** $j \leftarrow 1$ to n **do**
11:       $I_{big} \leftarrow \neg I_{small}[j, :]$
12:       $\hat{w}[j, I_{big}] \leftarrow \dot{X}[j, :]\Psi[I_{big}, :]^\dagger$
13:    **end for**
14: **end while**

Clearly $\Phi(x)\Phi^T(x)$ is a positive semidefinite matrix and $\gamma > 0$. Hence, this equation does not guarantee exponential convergence since $\Phi(x)$ may become zero, while convergence is not finished, i.e. when $w_e \neq 0$. Therefore, an extra condition is required to assure $\Phi(x)$ remains nonzero for some span of time so that the estimation error converges to zero. This is known as the Persistence of Excitation (PE) condition in adaptive control (see, e.g. Nguyen [2018]). Accordingly, it can be shown that if

$$\int_t^{t+T} \Phi(x)\Phi^T(x)\mathrm{d}\tau \geq \alpha_0 I,$$

is satisfied for all $t \geq t_0$ and some $\alpha_0 \geq 0$, then the asymptotic convergence to the optimal parameters is guaranteed.

4.6 Python Toolbox

Having implemented the model update algorithms discussed, we developed a Python toolbox that can be accessed at Farsi and Liu [2022].

In this toolbox, one has the option to choose from various techniques with different settings to draw a comparison among them. The algorithms can be compared in terms of the accuracy of the model and their computational complexity. Moreover, the comparison results are visualized using Seaborn [Waskom, 2021], a Python data visualization library.

In Section 4.6.1, we introduce the main features of this toolbox.

4.6.1 Configurations

To set up the simulations, we start by choosing different configurations of the number of samples, set of bases, learning algorithms for different systems. Moreover, we choose different type of plots for visualization of the results that includes the Root Mean Square Error (RMSE), weights, runtime, etc. Then, we iterate among different configurations of the systems and record the results to be compared later on. This is illustrated with the following Python script:

```
# Choose the learning algorithms by setting to 1.
select_ID_algorithm = {'SINDy': 1, 'RLS': 0,
    'Gradient Descent': 0, 'Least Squares':1}
# Choose different plots by setting to 1.
select_output_graphs={'Error':0,'Runtime':0,
    'Weights':0,'Residuals':0,
    'Runtime vs. num of Bases':0,
    'Runtime vs. num of Samples':1}
#Choose different set of bases to be compared.
List_of_chosen_bases = [['1', 'x', 'x^2', 'x^3'],
    ['1', 'x', 'sinx', 'cosx'],
    ['1', 'x', 'x^2','xx']]
#Choose different set of samples to be compared.
List_of_db_dim = [500,2000,4000]
List_of_systems=[Vehicle,Pendulum,Quadrotor,Lorenz]
#Use 4 nested loops to iterate among
#all configurations of systems.
for db_dim in List_of_db_dim:
 for index,Model in enumerate(List_of_systems):
  for chosen_bases in List_of_chosen_bases:
   for id in [i for i in select_ID_algorithm.keys()
            if select_ID_algorithm[i]==1]:
    ("...the main body of code...")
```

4.6.2 Model Update

To define and update the parameters of the model, we developed three Python classes: Library, Database, and SysID. We use Library to handle the operations regarding the bases, such as evaluation of the vector of bases and their partial derivatives. Then, in the case the learning algorithm requires a history of data to be saved over time, `Database` can be employed. Moreover, in `SysID`, we implement the model update algorithms discussed in Section 4.4.

These classes will be explained in detail in Chapter 10 as part of the Structured Online Learning (SOL) toolbox.

4.6.3 Model Validation

To analyze the model learned in terms of accuracy, one's first choice is to make predictions using the model obtained and compare them with the outputs from the original system. Therefore, for a given number of samples, we sample the system randomly at different points in the state and control spaces. Usually, different portions of this data are utilized to train and test the model. This procedure is well known as cross validation, where we only use the train data to fit a model. Then, the test data are used to measure the performance of the model. The main reason behind dividing data in two portions is to capture the performance against an independent set of data which are not seen while learning the model. This simulates the fact that in practice, predictions are made for new samples to which we do not have prior access.

Data Preparation
The training and test datasets are generated as follows:

```
#Randomly, sample the space of states and controls.
X = np.random.uniform(Domain[::2],
        Domain[1::2], size=(db_dim, n))
U = np.random.uniform(u_lim[::2],
        u_lim[1::2], size=(db_dim, m))
#Evaluate the dynamics.
for i in range(db_dim):
    X_dot[i,:]=Model.dynamics(0,X[i,:],U[i,:])
#Set the portion of training data.
#chosen from the dataset.
frac=0.8
#Separate the data into test and training sets.
X_dot_train,X_dot_test=X_dot[:int(frac*db_dim),:],
        X_dot[int(frac*db_dim):,:]
X_train,X_test=X[:int(frac*db_dim),:],
        X[int(frac*db_dim):,:]
U_train,U_test=U[:int(frac*db_dim),:],
        U[int(frac*db_dim):,:]
```

RMSE Computation

To compute the error, we consider the RMSE defined as follows:

$$E_{\mathrm{RMSE}} = \sqrt{\frac{1}{nN_s} \sum_{i=1}^{N_s} e_i^T e_i,} \tag{4.14}$$

where we also divide by n to get a uniform average of the error among all the dimensions of the dynamics.

Accordingly, we have the following block of code:

```
for i in range(len(X_train)):
    (...)
    #Upadate the model with a single training data.
    W,runtime[i,:] = SysID.update(X_train[i, :],
                      X_dot_train[i, :], U_train[i, :])
    #Compute RMSE for the current model
    for j in range(len(X_test)):
        sum+=np.sum(np.power(X_dot_test[j,:]-
                SysID.evaluate(X_test[j, :],
                U_test[j, :]),2))
    #Keep a record of error to be plotted later on.
    error_hist[i,:]=np.sqrt(sum/(n*len(X_test)))
```

It should be noted that the outer loop iterates for all training samples. This means we recalculate the RMSE whenever the model is updated by a new training sample. In this way, we capture the changes in the error while learning.

4.7 Comparison Results

In what follows, we review some benchmarking results generated by the toolbox presented in Section 4.7. It should be noted that making a fair comparison is not trivial due to different factors involved, such as the features of the system of interest, the data available, and the model chosen. Accordingly, here, we only run a general comparison to highlight the main characteristics of the algorithms. However, using the proposed tool, one can run simulations in different configurations to obtain more results for particular circumstances.

In the numerical results, we analyze the performance of the algorithms on two nonlinear dynamics including the Lorenz system (5.24) and the vehicle system (7.25). The definitions and details of these systems can be found in Sections 5.5 and 7.6.2, respectively.

In the following sections, we compare and analyze the parameter convergence, errors, and runtime results by choosing various models for learning of different systems.

4.7.1 Convergence of Parameters

In this part of the simulations, to investigate the parameter convergence, we implemented GD, RLS, LS, and Sparse Identification of Nonlinear Dynamics (SINDy) algorithms on the Lorenz system (5.24) given in Section 5.5.

We consider two scenarios to demonstrate the effect of bases chosen for the model, i.e. known and unknown bases. In the first case, we assume that the bases of the system are known. Hence, they can be included in the model. In this case, an exact model of the system can be potentially learned. In another case, we assume the true bases are not known. Hence, we include various bases in the model to obtain an approximation of the system. The evaluations of the parameters are demonstrated in Figures 4.2 and 4.3 for these two different scenarios.

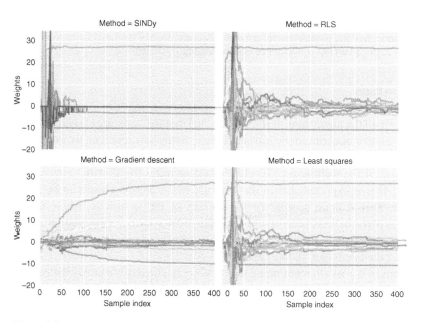

Figure 4.2 We assume the exact bases for the model are not know. Hence, only some of them may be present in the vector of bases chosen. Therefore, the model is approximated by various bases. The evolutions of the parameters for different algorithms are shown while learning.

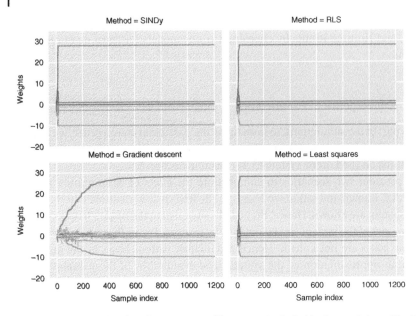

Figure 4.3 Assuming that the exact set of bases are included in the model, we illustrate the evolutions of the parameters for different algorithms while learning. Compared to Figure 4.2, the quality of convergence is considerably enhanced.

Unknown Bases

In Figure 4.2, one can observe that the parameters approach to almost the same results for all the algorithms. It should be noted that the bases are chosen in a way that an exact model cannot be learned. Hence, in the best scenario, we aim on an approximate model. Therefore, the parameters are not uniquely estimated.

Among all the techniques, GD clearly demonstrates the slowest convergence results. This is expected due to the learning rate which is chosen to be small enough. This is done because of the stability of the technique, where GD is prone to perturbations because only a single point is used for calculation of the gradients. This can be improved by using the normalized version of this algorithm to adjust the parameter update gain.

Moreover, it is worth mentioning that SINDy can efficiently sparsify the parameters as expected. It can be observed that the small values are eliminated over time, while other techniques estimate many parameters with small values.

Known Bases

In Figure 4.3, we demonstrate the parameter convergence for the case the bases are known. As seen in this scenario, SINDy can discover the exact dynamics by sparsification. In a similar result, GD and RLS can obtain an almost accurate model, while unlike to SINDy, they include many small parameters that impede obtaining the exact analytical model of the system. Similar to the results in Figure 4.2, GD shows oscillations which prevent convergence to a more accurate model.

4.7.2 Error Analysis

Unknown Bases

For all the four algorithms, we draw the RMSE graphs in Figure 4.4, where the computation of these errors are explained earlier in section "RMSE Computation". It is observed that except GD, the rest of the techniques converge to a similar error. However, GD experience difficulties in minimizing the error due to the oscillations.

Furthermore, we illustrate the residuals for the Lorenz system in terms of the states and the control in Figure 4.5, where e_i denotes the residual for the ith dynamic. It can be seen that in the first dimension, the model is efficiently trained. However, in the second and third dimensions, the model learned is not accurate enough. This is expected regarding the fact that the chosen bases do not include all the terms of the Lorenz dynamic.

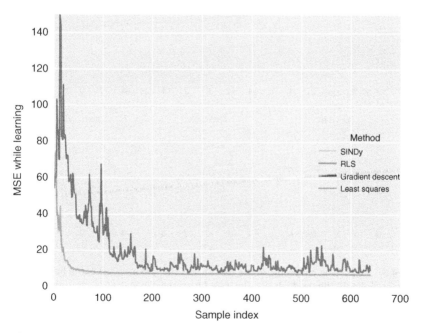

Figure 4.4 A comparison of the RMSE for different algorithms while learning. In this comparison, the exact bases are only partially included in the model.

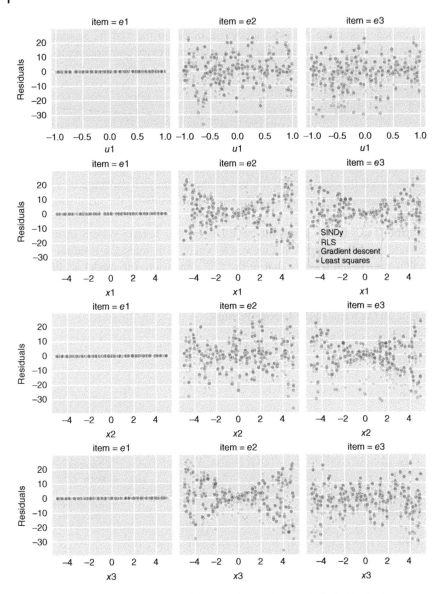

Figure 4.5 We assume only some of the exact bases that are included in the bases chosen. Accordingly, the residuals are illustrated for three dimensions of the Lorenz system in terms of the states and the control, where e_i denotes the residual for the ith component of the dynamics.

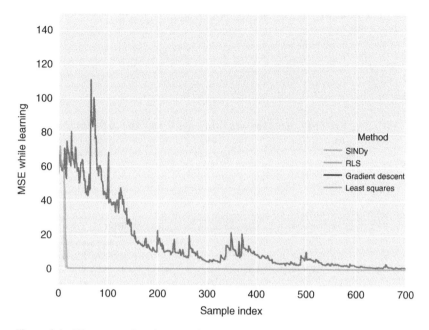

Figure 4.6 We assume that the exact bases are known, and they are included in the model. A comparison of the RMSE for different algorithms while learning.

Known Bases

In a similar way as Section 4.7.2.1, we draw a comparison among RMSE of all four algorithms in Figure 4.6, where we assume the bases are known. It is evident that except GD, all the other techniques can quickly identify the system dynamics with almost zero error.

To investigate the models in more detail, consider the residuals shown in Figure 4.7. It can be verified that using the exact set of bases, the models obtained are considerably more accurate than the case with unknown bases shown in Figure 4.5. According to Figure 4.7, all three algorithms SINDy, GD, and RLS are able to converge to the exact system since the residuals are almost zero. On the other hand, GD should be further improved in the first and second dynamic, where both show considerable variance. Moreover, the model is biased in second dynamic. One way to improve the model is to decrease the learning rate, and instead, increase the number of samples used for updating the model.

4.7.3 Runtime Results

We draw a comparison among all the techniques in terms of the complexity of the computations involved in Figure 4.8. This is done by plotting the Kernel Density

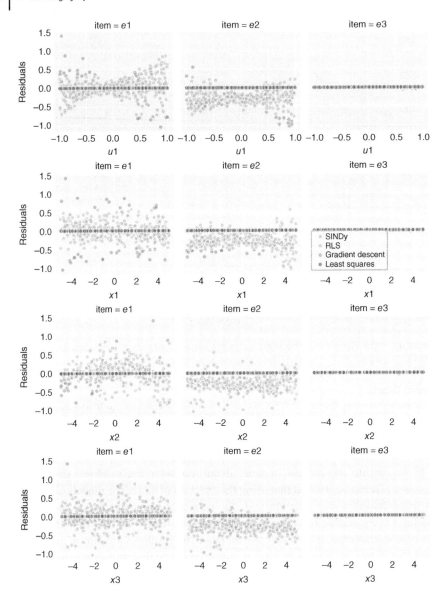

Figure 4.7 Assuming that the exact bases are included in the set of bases chosen, we demonstrate the residuals obtained for three dimensions of the Lorenz system in terms of the states and the control, where e_i denotes the residual for the ith component of the dynamics.

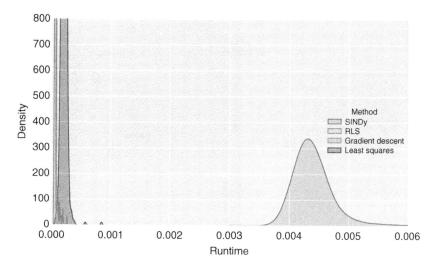

Figure 4.8   A comparison of runtimes for different algorithms.

Estimation (KDE) of the runtimes for updates in the model. It is evident that GD and RLS are considerably faster than two other techniques. This is consistent with the fact that these two techniques employ only one sample at a time to update the model, while LS and SINDy use a growing batch of recorded data. Moreover, SINDy clearly requires the longest time to update the parameters. This is expected given the extra iterations involved in this algorithm, compared with LS.

Increasing the Number of Bases

In Figure 4.9, by increasing the number of bases for Lorenz and vehicle systems, with $n = 3$ and $n = 5$, respectively, we analyze the changes in the runtime results. It is observed that GD and RLS scale well with increasing number of bases, while the runtime considerably increases for SINDy and LS. To provide more clear comparison between GD and RLS, separate results are presented in Figure 4.10. It is evident that GD can run only slightly faster than RLS when a limited number of bases are used. However, if one chooses to use large number of bases, the gap tends to be larger.

Increasing the Number of Samples

Moreover, to analyze whether the approaches can deal with large number of samples, we perform another simulation by considerably increasing number of samples at each iteration. Considering that RLS and GD use only the current sample to update the model, their computational cost is not determined by the number of samples. Therefore, we only consider SINDy and LS for which the computations are directly affected with the number of samples. In Figure 4.11, one can verify this conclusion.

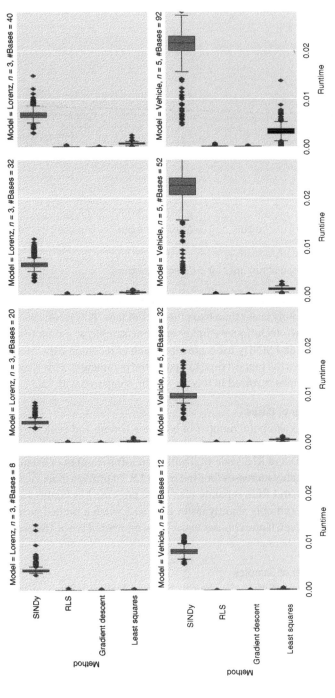

Figure 4.9 The effect of increasing number of bases on the runtime of different algorithms is investigated for the Lorenz and vehicle system.

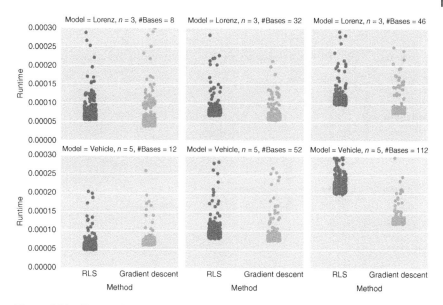

Figure 4.10 The runtime of GD and RLS are compared in terms of the increasing number of bases, for the Lorenz and vehicle system.

4.8 Summary

In this chapter, we considered data and model as two main components in learning of dynamical systems. We formulated the learning of control systems according to samples that can be obtained from the state and control to update a model selected. Then, we presented different algorithms for updating the model and discuss their characteristics.

According to our prior knowledge about the system, the learning can be formulated in gray-box or black-box fashions. If the assumptions of the problem allows employing a gray-box model, learning can be done more efficiently compared to black-box case. However, this option is not often available. Hence, taking a black-box approach is inevitable.

The unknown system can be approximated using parametric and nonparametric models. Although these techniques provide more flexibility in the applications, specific considerations should be taken into account depending on the model. Their efficiency is usually determined in a trade-off between the simplicity and accuracy of the model. In this regard, parametric models can scale efficiently with the number of samples and can adapt to different nonlinearities. Therefore, we consider a specific family of parametric models which linearly depend on their parameters to learn dynamical systems.

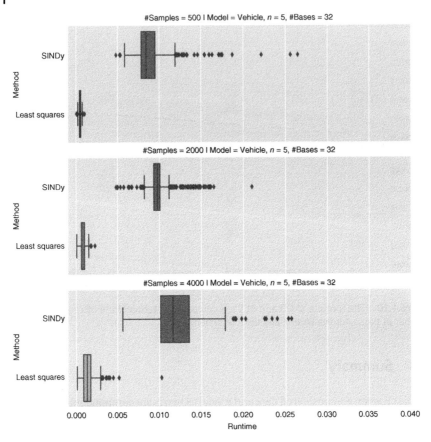

Figure 4.11 The runtime of SINDy and LS are compared in terms of the increasing number of samples, for the Lorenz and vehicle system.

Assuming that samples of the system are available, different algorithms can be found for learning linearly parameterized models. LS and SINDy were discussed as two parameter update methods that are applied in the batch mode. These techniques can effectively learn parameters, where SINDy is specifically designed to discover the underlying dynamics of a class of system. Due to the fact that the complexity of these techniques increases considerably with the number of samples, they normally are implemented offline. However, for less-complex systems, these techniques may be still applicable online by limiting the samples to a short list that can abstract out the dynamics. According to the simulations, the complexity of RLS and GD, as online techniques, does not depend on the number of samples. Compared to RLS, GD is more computationally efficient, while RLS demonstrated more stable results.

Bibliography

Maurice S. Bartlett. An inverse matrix adjustment arising in discriminant analysis. *The Annals of Mathematical Statistics*, 22(1):107–111, 1951.

Steven L. Brunton, Joshua L. Proctor, and J. Nathan Kutz. Discovering governing equations from data by sparse identification of nonlinear dynamical systems. *Proceedings of the National Academy of Sciences of the United States of America*, 113(15):3932–3937, 2016.

Milad Farsi and Jun Liu. Python toolbox for structured online learning-based (SOL) control. https://github.com/Miiilad/SOL-Python-toolbox, 2022. [Online; accessed 4-June-2022].

Karel J. Keesman. *System Identification: An Introduction*. Springer, 2011.

Jianping Li. General explicit difference formulas for numerical differentiation. *Journal of Computational and Applied Mathematics*, 183(1):29–52, 2005.

Lennart Ljung and Torsten Söderström. *Theory and Practice of Recursive Identification*. MIT Press, 1983.

Nhan T. Nguyen. Model-reference adaptive control. In *Model-Reference Adaptive Control*, Michael J. Grimble, Michael A. Johnson, and Linda Bushnell (eds.), pages 83–123. Springer, 2018.

Jack Sherman. Adjustment of an inverse matrix corresponding to changes in the elements of a given column or a given row of the original matrix. *Annals of Mathematical Statistics*, 20(4):621, 1949.

Michael L. Waskom. Seaborn: Statistical data visualization. *Journal of Open Source Software*, 6(60):3021, 2021.

5

Structured Online Learning-Based Control of Continuous-Time Nonlinear Systems

5.1 Introduction

In this chapter, we introduce a Model-based Reinforcement Learning (MBRL) technique for control of nonlinear continuous-time systems with unknown dynamics. We focus on stabilization of an equilibrium point in this chapter. In Section 5.2, we formulate an optimal control approach based on a particular structure of dynamics and characterize the optimal feedback control based on a matrix of parameters obtained by a differential equation. Section 5.4 outlines the Structured Online Learning (SOL) algorithm designed based on the obtained results. In Section 5.5, we present the numerical results of this algorithm implemented on a few benchmark examples. The contents of this chapter are published in Farsi and Liu [2020].

5.2 A Structured Approximate Optimal Control Framework

Consider the nonlinear control-affine system

$$\dot{x} = F(x, u) = f(x) + g(x)u, \tag{5.1}$$

where $x \in D \subseteq \mathbb{R}^n$ and $u \in \Omega \subseteq \mathbb{R}^m$ are the state and the control input, respectively. Moreover, $f : D \to \mathbb{R}^n$ and $g : D \to \mathbb{R}^{n \times m}$.

The cost functional to be minimized along the trajectory, starting from the initial condition $x_0 = x(0)$, is considered in the following linear quadratic form:

$$J(x_0, u) = \lim_{T \to \infty} \int_0^T e^{-\gamma t} \left(x^T Q x + u^T R u \right) dt, \tag{5.2}$$

where $Q \in \mathbb{R}^{n \times n}$ is positive semidefinite, $\gamma \geq 0$ is the discount factor, and $R \in \mathbb{R}^{m \times m}$ is a diagonal matrix with only entries of positive values, given by design

Model-Based Reinforcement Learning: From Data to Continuous Actions with a Python-based Toolbox, First Edition. Milad Farsi and Jun Liu.

criteria. With $\gamma > 0$, this defines a discounted optimal control problem, which has been discussed, e.g. in Postoyan et al. [2014] and Gaitsgory et al. [2015]. Moreover, it is widely used in Reinforcement Learning (RL) to determine the time horizon considered for minimizing the objective [Lewis and Liu, 2013].

For the closed-loop system, by assuming a feedback control law $u = \omega(x(t))$ for $t \in [0, \infty)$, the optimal control is given by

$$\omega^* = \arg \min_{u(\cdot) \in \Gamma(x_0)} J(x_0, u(\cdot)), \tag{5.3}$$

where Γ is the set of admissible controls.

Assumption 5.1 *Each component of f and g can be identified or effectively approximated within the domain of interest by a linear combination of some basis functions $\phi_i \in C^1 : D \to \mathbb{R}$ for $i = 1, 2, \ldots, p$.*

Accordingly, (5.1) is rewritten as follows:

$$\dot{x} = W\Phi(x) + \sum_{j=1}^{m} W_j \Phi(x) u_j, \tag{5.4}$$

where W and $W_j \in \mathbb{R}^{n \times p}$ are the matrices of the coefficients obtained for $j = 1, 2, \ldots, m$, and $\Phi(x) = [\phi_1(x) \ \ldots \ \phi_p(x)]^T$. Moreover, the bases are chosen such that the first elements include the constant and linear terms as $\Phi = [1 \ x_1 \ \ldots \ x_n \ \phi_{n+2}(x) \ \ldots \ \phi_p(x)]^T$

In what follows, without loss of generality, the cost defined in (5.2) is transformed to the space of bases $\Phi(x)$ as follows:

$$J(x_0, u) = \lim_{T \to \infty} \int_0^T e^{-\gamma t} \left(\Phi(x)^T \bar{Q} \Phi(x) + u^T R u \right) dt, \tag{5.5}$$

where $\bar{Q} = \text{diag} \left(0, Q, \mathbf{0}_{(p-n-1) \times (p-n-1)} \right)$ is a block diagonal matrix that contains all zeros except the second block Q which corresponds to the linear bases x, following from Φ in Assumption 5.1.

Then the corresponding Hamilton–Jacobi–Bellman (HJB) equation can be written as follows:

$$-\frac{\partial}{\partial t}(e^{-\gamma t} V) = \min_{u(\cdot) \in \Gamma(x_0)} H, \tag{5.6}$$

where the Hamiltonian H is defined as follows:

$$H = e^{-\gamma t} \left(\Phi(x)^T \bar{Q} \Phi(x) + u^T R u \right) + e^{-\gamma t} \frac{\partial V}{\partial x} \left(W\Phi(x) + \sum_{j=1}^{m} W_j \Phi(x) u_j \right). \tag{5.7}$$

In general, there exists no analytical approach that can solve such a partial differential equation and obtain the optimal value function. However, it has been shown in the literature that approximate solutions can be computed by numerical techniques [Wang et al., 2009; Lewis and Vrabie, 2009; Balakrishnan et al., 2008; Kharroubi et al., 2014; McEneaney, 2007; Kang and Wilcox, 2017].

We assume a parameterization of the optimal value function in the following form:

$$V = \Phi(x)^T P \Phi(x), \tag{5.8}$$

where $P \in \mathbb{R}^{p \times p}$ is symmetric.

Remark 5.1 Unlike other approximate optimal approaches in the literature, such as Zhang et al. [2011], Bhasin et al. [2013], and Kamalapurkar et al. [2016a], which use a linear combination of bases to parameterize the value function, we assume a quadratic form. As a result, the value function now is defined in the product space $\Lambda := \Phi \times \Phi$. Hence, it is expected that the resulting quadratic terms better contribute to basing a positive value function around $x = 0$. Furthermore, due to the function-approximating properties of the basis vector Φ itself, one may bring them to Λ in addition by including a constant basis c in Φ. Therefore, compared to other approaches, the structure used in (5.8) suggests a more compact way of formulating the problem, where by only a limited number of bases in Φ, we can attain a richer set Λ to parameterize the value function.

Then the Hamiltonian is given by

$$H = e^{-\gamma t}(\Phi(x)^T \bar{Q} \Phi(x) + u^T R u)$$

$$+ e^{-\gamma t} \Phi(x)^T P \frac{\partial \Phi(x)}{\partial x} \left(W \Phi(x) + \sum_{j=1}^{m} W_j \Phi(x) u_j \right)$$

$$+ e^{-\gamma t} \left(\Phi(x)^T W^T + \sum_{j=1}^{m} u_j^T \Phi(x)^T W_j^T \right) \frac{\partial \Phi(x)}{\partial x}^T P \Phi(x).$$

Moreover, based on the structure of R, the quadratic term of u is rewritten in terms of its components:

$$H = e^{-\gamma t} \left(\Phi(x)^T \bar{Q} \Phi(x) + \sum_{j=1}^{m} r_j u_j^2 + \Phi(x)^T P \frac{\partial \Phi(x)}{\partial x} W \Phi(x) + \right.$$

$$\Phi(x)^T P \frac{\partial \Phi(x)}{\partial x} \left(\sum_{j=1}^{m} W_j \Phi(x) u_j \right) + \Phi(x)^T W^T \frac{\partial \Phi(x)}{\partial x}^T P \Phi(x)$$

$$\left. + \left(\sum_{j=1}^{m} u_j \Phi(x)^T W_j^T \right) \frac{\partial \Phi(x)}{\partial x}^T P \Phi(x) \right), \tag{5.9}$$

where $r_j \neq 0$ is the jth component on the diagonal of matrix R. To minimize the resulting Hamiltonian, we need

$$\frac{\partial H}{\partial u_j} = 2 r_j u_j + 2 \Phi(x)^T P \frac{\partial \Phi(x)}{\partial x} W_j \Phi(x) \tag{5.10}$$

$$= 0, \quad j = 1, 2, \dots, m.$$

Hence, the jth optimal control input is obtained as follows:

$$u_j^* = -\Phi(x)^T r_j^{-1} P \frac{\partial \Phi(x)}{\partial x} W_j \Phi(x). \tag{5.11}$$

By plugging in the optimal control and the value function in (5.6), we get

$$- e^{-\gamma t} \Phi(x)^T \dot{P} \Phi(x) + \gamma e^{-\gamma t} \Phi(x)^T P \Phi(x)$$

$$= e^{-\gamma t} \left(\Phi(x)^T \bar{Q} \Phi(x) \right.$$

$$+ \Phi(x)^T P \frac{\partial \Phi(x)}{\partial x} \left(\sum_{j=1}^{m} W_j \Phi(x) r_j^{-1} \Phi(x)^T W_j^T \right) \frac{\partial \Phi(x)}{\partial x}^T P \Phi(x)$$

$$- 2\Phi(x)^T P \frac{\partial \Phi(x)}{\partial x} \left(\sum_{j=1}^{m} W_j \Phi(x) r_j^{-1} \Phi(x)^T W_j^T \right) \frac{\partial \Phi(x)}{\partial x}^T P \Phi(x)$$

$$+ \Phi(x)^T P \frac{\partial \Phi(x)}{\partial x} W \Phi(x) + \Phi(x)^T W^T \frac{\partial \Phi(x)}{\partial x}^T P \Phi(x) \right).$$

This is rewritten as follows:

$$- \Phi^T \dot{P} \Phi(x) + \gamma \Phi(x)^T P \Phi(x) = \Phi(x)^T \bar{Q} \Phi(x)$$

$$- \Phi(x)^T P \frac{\partial \Phi(x)}{\partial x} \left(\sum_{j=1}^{m} W_j \Phi(x) r_j^{-1} \Phi(x)^T W_j^T \right) \frac{\partial \Phi(x)}{\partial x}^T P \Phi(x)$$

$$+ \Phi(x)^T P \frac{\partial \Phi(x)}{\partial x} W \Phi(x) + \Phi(x)^T W^T \frac{\partial \Phi(x)}{\partial x}^T P \Phi(x), \tag{5.12}$$

where a sufficient condition to hold this equation is

$$-\dot{P} = \bar{Q} + P \frac{\partial \Phi(x)}{\partial x} W + W^T \frac{\partial \Phi(x)}{\partial x}^T P - \gamma P$$

$$- P \frac{\partial \Phi(x)}{\partial x} \left(\sum_{j=1}^{m} W_j \Phi(x) r_j^{-1} \Phi(x)^T W_j^T \right) \frac{\partial \Phi(x)}{\partial x}^T P. \tag{5.13}$$

This equation has to be solved backward to get a value of P that characterizes the optimal value function (5.8) and control (5.11). However, it has been shown that the forward integration of such an equation converges to similar solutions, as long as we are not very close to the initial time, i.e. in the steady-state mode (Prach et al. [2015]; see also Theorem 2.2).

Remark 5.2 While the similarity between the derived optimal control and the Linear Quadratic Regulator (LQR) problem cannot be denied, there are substantial differences. Notably, the matrix differential equation (5.13) is derived in terms of Φ, which is of dimension p in contrast with the LQR formulation that includes only the linear terms of the state with dimension n.

Remark 5.3 Because of the general case considered in obtaining (5.13), where Φ includes arbitrary basis functions of the state, there seems no way to escape from the state-dependency in this equation, except in the linear case as mentioned. Hence, we require (5.13) be solved along the trajectories of the system.

5.3 Local Stability and Optimality Analysis

In this section, we will present the stability analysis of the approach and its connections with the Forward-Propagating Riccati Equation (FPRE) for linear systems [Weiss et al., 2012; Prach et al., 2015] as discussed in Section 2.3.4. To do this, we first formulate the LQR problem for the linearized model of (5.4). Then we will show that, once we get close enough to the origin, the integration of (5.13) will be governed by the forward-propagated solution of the linearized system, as the dominant part.

Before we start, we need some reformulation on the system that can be assured with no loss of generality. Consider the structured system (5.4). We will assume an equilibrium point at the origin. Moreover, we need to redefine the bases by using the following lemma:

Lemma 5.1 *Assuming that the constant basis $\phi_1(x) = 1$ is included in Φ, we can always redefine Φ such that $\phi_i(0) = 0$ for $i = 2, \ldots, p$. To hold these properties, we redefine $\phi_i(x) := \phi_i(x) - \phi_i(0)$ for $i = 2, \ldots, p$. Accordingly, we also set the W to zeros in the entries corresponding to basis 1. For instance, the system $\dot{x} = 1 - \cos x = [1 - 1]\begin{bmatrix} 1 \\ \cos x \end{bmatrix}$ can be equivalently rewritten as $\dot{x} = [0 - 1]\begin{bmatrix} 1 \\ \cos x - 1 \end{bmatrix}$.*

Accordingly, we construct the vector of basis as $\Phi = [1 \quad x^T \quad \Gamma(x)^T]^T$, where Γ includes all the nonlinear bases. Then, according to Lemma 5.1, the system (5.4) is represented using some block-structured matrices as follows:

$$\dot{x} = \begin{bmatrix} 0 & W_2 & W_3 \end{bmatrix}\begin{bmatrix} 1 \\ x \\ \Gamma(x) \end{bmatrix} + \begin{bmatrix} W_{j_1} & W_{j_2} & W_{j_3} \end{bmatrix}\begin{bmatrix} 1 \\ x \\ \Gamma(x) \end{bmatrix}u. \tag{5.14}$$

5.3.1 Linear Quadratic Regulator

By defining $\Gamma_1 = \frac{\partial \Gamma(x)}{\partial x}|_{x=0}$, the linearization of (5.14) at the equilibrium point yields

$$\dot{x} = Ax + \sum_{j=1}^{m} B_j u_j, \tag{5.15}$$

where $A = W_2 + W_3\Gamma_1$ and $B_j = W_{j_1}$.

Now, consider the LQR problem with quadratic cost (5.2) and $\gamma = 0$ for the linearized system (5.15). Then, the optimal control is given by $u_j = -r_j^{-1}B_j^T\bar{S}x$, where $\bar{S}$ is the solution of the well-known Algebraic Riccati Equation (ARE):

$$Q + \bar{S}A + A^T\bar{S} - \bar{S}\left(\sum_{j=1}^{m}B_jr_j^{-1}B_j^T\right)\bar{S} = 0. \tag{5.16}$$

Alternatively, we consider the forward solution of the following Differential Riccati Equation (DRE), where we update the feedback controller with the solution $S(t)$ at any $t \in [0, \infty)$ using

$$u_j = -r_j^{-1}B_j^TSx,$$

$$\dot{S} = Q + SA + A^TS - S\left(\sum_{j=1}^{m}B_jr_j^{-1}B_j^T\right)^T S. \tag{5.17}$$

By substitution of A and B, we get

$$u_j = -r_j^{-1}W_{j_1}^TSx, \tag{5.18}$$

$$\dot{S} = Q + S(W_2 + W_3\Gamma_1) + (W_2^T + \Gamma_1^TW_3^T)S$$
$$- S\left(\sum_{j=1}^{m}W_{j_1}r_j^{-1}W_{j_1}^T\right)S. \tag{5.19}$$

The following lemma is a restatement of Theorem 2.2, which is based on Theorems 1 and 4 in Prach et al. [2015], but strengthens the conclusion there (from asymptotic convergence to global uniform exponential stability) under slightly weaker assumptions (with (A, B) being stabilizable instead of controllable).

Lemma 5.2 *Assume that (A, B) is stabilizable and Q is positive definite. Consider plant (5.15) with the control law (5.17), where, for all $t \in [0, \infty)$, $S(t)$ is the positive semidefinite solution of the FPRE (5.17) with $S(0) \geq 0$. Then, the origin is Globally Uniformly Exponentially Stable (GUES) for the closed-loop system. Moreover, as $t \to \infty$, $S(t)$ will converge to $\bar{S}$.*

5.3.2 SOL Control

Based on the optimal control framework presented, for a known structured nonlinear system in the form of (5.14), the optimal control is given by (5.11). Moreover, the value is updated by the evolutions of the parameters in (5.13). Accordingly, we make the following remarks on guarantees for the stability of the closed-loop system.

Remark 5.4 Let $\bar{D}_r = \{x \in D : \| x \| < r\}$. Suppose that the solution $P(t)$ of (5.13) starting from $P(0) = 0$ establishes a controller such that solutions of the closed-loop system remain (or asymptotically remain) in D_r. We can perform an asymptotic analysis to show that $P(t)$ will lead to a stabilizing controller the nonlinear closed-loop system. See Appendix 11.1 for more details. However, it seems challenging to theoretically establish such a controlled invariance property through online computation of $P(t)$. In the proof of Lemma 5.2, the explicit solution of the FPRE are used. However, to analyze its performance on the nonlinear system, we need to analyze a perturbed version of the FPRE, which seems fundamentally more challenging. This challenge is also expected, because in the RL approaches (mainly Policy Iteration (PI) algorithms) we introduced in Chapter 3, an initial stabilizing policy is always assumed. Here, we do not assume such an initial policy.

Assuming that we can show asymptotic stability of the closed-loop system in some region around the equilibrium point, we further make the following remark on optimality of the controller.

Remark 5.5 Assume that the conditions of Remark 5.4 holds and a local stabilizing controller is obtained, with a sufficiently small choice of the discounting factor, $\gamma \to 0$. Then, the feedback control rule converges to the LQR control of the linearized system given by (5.16). Hence, the local optimality of the obtained controller is asymptotically guaranteed. See Appendix 11.2 for more details.

In Section 5.4, we will establish an online learning algorithm based on the proposed optimal control framework.

5.4 SOL Algorithm

By considering a general description of the nonlinear input-affine system in terms of some bases as in (5.4), we obtained a structured optimal control framework that suggests using the state-dependent matrix differential equation (5.13) to achieve the parameters of the nonlinear feedback control. Next, we exploit this framework to propose the SOL algorithm. Hence, the focus of this section will be on the algorithm and practical properties of SOL.

The learning procedure is done in the following order: First, initialize $P(0)$ with a zero matrix. Then, in the control loop:

- We acquire the samples of the states at any time step t_k and evaluate the set of bases accordingly.

- We update the structured system model by a system identification technique.
- Using the measurements and the updated model coefficients, we integrate (5.13) to update $P(t_k)$.
- We calculate the control value using (5.11) for the next step t_{k+1} using $P(t_k)$.
- $k + +$.

In what follows, we discuss the steps involved in more details by focusing on the Sparse Identification of Nonlinear Dynamics (SINDy) algorithm for identification.

5.4.1 ODE Solver and Control Update

In this approach, we run the system from some $x_0 \in D$, then solve the matrix differential equation (5.13) along the trajectories of the system. Different solvers are already developed that can efficiently integrate differential equations. In the simulation, we use a Runge–Kutta solver to integrate the dynamics of the system. In real-world applications, this can be replaced with an approximation obtained from measurements of the real system states. Although the solver may take smaller steps, we only allow the measurements and control update at time steps $t_k = kh$, where h is the sampling time and $k = 0, 1, 2, \dots$. For solving (5.13) in continuous time, we use the Runge–Kutta solver with a similar setting, where the weights and the states in this equation are updated by a system identification algorithm and the measurements x_k at each iteration of the control loop, respectively. A recommended choice for P_0 is a matrix with components of zero or very small values.

The differential equation (5.13) also requires evaluations of $\partial \Phi / \partial x_k$ at any time step. Since the bases Φ are chosen beforehand, the partial derivatives can be analytically calculated and stored as functions. Hence, they can be evaluated for any x_k in a similar way as Φ itself. By solving (5.13), we can calculate the control update at any time step t_k according to (5.11). Although, at the very first steps of learning, control is not expected to take effective steps toward the control objective, it can help in exploration of the state space and gradually improve by learning more about the dynamics.

Remark 5.6 The computational complexity of updating parameters by relation (5.13) is bounded by the complexity of matrix multiplications of dimension p which is $\mathcal{O}(p^3)$. Moreover, it should be noted that, regarding the symmetry in the matrix of parameters P, this equation updates $L = (p^2 + p)/2$ number of parameters which correspond to the number of bases used in the value function. Therefore, in terms of the number of parameters, the complexity of the proposed technique is $\mathcal{O}(L^{3/2})$. However, for instance, if a Recursive Least Squares (RLS) technique was employed with the same number of parameters, the computations are bounded by $\mathcal{O}(L^3)$. As a result, the proposed parameter update scheme can be done considerably faster than similar model-based techniques, such as Kamala-purkar et al. [2016b] and Bhasin et al. [2013]. In another effort, Kamalapurkar et al. [2016a] decreased the number of bases used to improve the computational efficiency, while the complexity still remained as $\mathcal{O}(L^3)$.

5.4.2 Identified Model Update

We considered a given structured nonlinear system as in Assumption 5.1. There-fore, having the control and state samples of the system, we need an algorithm that updates the estimation of system weights. As studied in Brunton et al. [2016] and Kaiser et al. [2018], SINDy is a data-efficient tool to extract the underlying sparse dynamics of the sampled data. Hence, we use SINDy to update the weights of the system to be learned. In this approach, along with the identification, the sparsity is also promoted in the weights by minimizing

$$[\hat{W} \ \hat{W}_1 \ \dots \ \hat{W}_m]_k = \arg\min_{\bar{W}} \ \| \dot{X}_k - \bar{W}\Theta_k \|_2^2 + \lambda \| \bar{W} \|_1, \tag{5.20}$$

where k is the time step, $\lambda > 0$, and Θ_k includes a matrix of samples with the columns of

$$\Theta_k^s = [\Phi^T(x^s) \ \Phi^T(x^s)u_1^s \ \dots \ \Phi^T(x^s)u_m^s]_k^T,$$

for sth sample. In the same order, $\dot{X}_k$ keeps a table of sampled state derivatives.

Updating $\hat{W}_k$ based on a history of samples may not be favored as the number of samples needed tends to be large. Especially, real-time implementations may not be possible because of the latency caused by the computations. There exist other techniques that can be alternatively used in different situations, such as neural networks, nonlinear regression, and other function approximation and system identification methods. For real-time control applications, considering the linear dependence on the system weights in (5.4), one may choose the RLS update rule that only uses the latest sample of the system and $\hat{W}_{k-1}$, which will run considerably faster.

5.4.3 Database Update

For using SINDy algorithm, a database of samples is required to recursively perform regressions at each time step. These weights correspond to a library of functions given in Φ. Any sample of the system at time k, to be stored in the database, includes Θ_k^s and the derivatives of the states approximated by $\hat{\dot{x}}_k = (x_k - x_{k-1})/h$.

For better results, higher-order approximations of the state derivative can be employed that may also include future samples of the states, for instance, x_{k+1}, x_{k+2}. To make this possible in the implementations, the identification should lag a few steps behind the controller update. The effect of lagging for a few steps is minor and can be safely neglected in the long run.

We adopt SINDy to do an online learning task, meaning that the database has to be gradually built along with the exploration and control. Different approaches can be employed in the choice of samples and building a database online. A comparison of these techniques can be found in Kivinen et al. [2004] and Van Vaerenbergh and Santamaría [2014].

In the implementations of SOL done in this chapter, we assume a given maximum size of database N_d, then we keep adding the samples with larger prediction errors to the database. Therefore, at any step we compare the prediction error $\dot{e}_k = \| \dot{x}_k - \hat{\dot{x}}_k \|$ with the average $\bar{\dot{e}}_k = \sum_{i=1}^{k} \dot{e}_k / k$. Hence, if the condition $\dot{e}_k > \eta \bar{\dot{e}}_k$ holds, we add the sample to the database, where the constant $\eta > 0$ adjusts the threshold. Choosing smaller values of η will increase the rate of adding samples to the database.

This procedure is done in a loop together with updating the control until a bound of the average prediction error is obtained that allows the controller to regulate the system to the given reference state. If the maximum number of samples in database is reached, we forget the oldest sample and replace it with the recent one. Hence, η should not be set too low to avoid fast forgetting of the older useful samples.

5.4.4 Limitations and Implementation Considerations

In this section, we discuss some considerations should be taken into account prior to running the algorithm:

- It should be noted that, although the learning approach is validated only in the simulation environment here, the SOL algorithm is proposed to be implemented on real-world problems. Hence, the training is meant to take place in real-time on real systems that requires computational and data efficiency of the algorithm.
- It is assumed that the environment can be made safe in a region of interest, and there exists a resetting mechanism if the system state reaches the boundary of the region. This allows trials and errors with no considerable damage to the system within a limited number of episodes until the stability can be preserved.
- Considering the control problem formulation, the control and state spaces, and the time horizon cannot be hardly constrained. However, they can be tuned by using the parameters specified by the user in the objective (5.2) including R, Q, and γ, respectively.
- Depending on the system identification technique implemented, there usually exist some tuning parameters. Having an initial knowledge of the system can help greatly in setting these parameters, as well as in choosing the set of bases.

5.4.5 Asymptotic Convergence with Approximate Dynamics

Consider the system structured as follows:

$$\dot{x} = W\Phi + \sum_{j=1}^{m} W_j \Phi \hat{u}_j + \epsilon. \tag{5.21}$$

where $\hat{u}_j = -\Phi^T r_j^{-1} \hat{P} \Phi_x \hat{W}_j \Phi$ is the feedback control rule obtained based on the estimation of the system $(\hat{W}, \hat{W}_j)$. Moreover, ϵ is the bounded approximation error in D. By assuming $W = \hat{W} + \tilde{W}$ and $W_j = \hat{W}_j + \tilde{W}_j$, this can be rewritten as follows:

$$\dot{x} = \hat{W}\Phi + \sum_{j=1}^{m} \hat{W}_j \Phi \hat{u}_j + \Delta(t), \tag{5.22}$$

where unidentified dynamics are lumped together as $\Delta(t)$. By the assumption that the feedback control u_j is bounded in D, we have $\| \Delta(t) \| \leq \bar{\Delta}$. For asymptotic convergence, and also to promote the robustness of the controller, the effect of the uncertainty should be taken into account. Hence, we use an auxiliary vector ρ to get

$$\dot{x} = \hat{W}\Phi + \sum_{j=1}^{m} \hat{W}_j \Phi \hat{u}_j + \Delta(t) + \rho - \rho$$

$$= \hat{W}_\rho \Phi + \sum_{j=1}^{m} \hat{W}_j \Phi \hat{u}_j + \Delta(t) - \rho,$$

where assuming that Φ also includes the constant basis, we adjusted the corresponding column in the system matrix to get $\hat{W}_\rho$. In the case $\bar{\Delta} = 0$, by using Remark 5.4, the controller $\hat{u}$ can be obtained such that the closed-loop system is locally asymptotically stable. For the case $\bar{\Delta} > 0$, although the system will stay stable for small enough $\bar{\Delta}$, it may not asymptotically converge to zero. Then, similar to Xian et al. [2004] and Qu and Xu [2002], we obtain ρ as below to help sliding the system state to zero

$$\rho = \int_0^t [k_1 x(\tau) + k_2 \operatorname{sign}(x(\tau))] d\tau,$$

where k_1 and k_2 are positive scalars. It can be shown that over time $\| \Delta(t) - \rho \| \to 0$, and hence, the system will asymptotically converge to the origin.

5.5 Simulation Results

We have implemented the proposed approach on four examples, which are presented in two categories considering Assumption 5.1: (i) the dynamics can be written exactly in terms of some choice of basis functions, and (ii) the dynamics include some terms that are required to be approximated in the space of some given bases.

As mentioned, in these numerical examples, we have exploited the SINDy algorithm for the identification purpose. However, clearly the focus of the simulations here is on the properties of the proposed control scheme rather than the identification part, considering that SINDy already has been extensively studied in Brunton et al. [2016] and Kaiser et al. [2018] as an offline identification algorithm. The SINDy algorithm adopted here is a powerful tool to obtain the dynamics of the system with a good precision. However, this depends greatly on how efficiently we can approximate the derivatives of the states. Hence, in different implementations, higher sampling rates or higher-order approximation of the derivatives may be needed. For the same reason, in the proposed examples, the number of samples used and the system obtained may be further tuned to match the level of quality reported in Brunton et al. [2016] and Kaiser et al. [2018].

The simulations are done in Python, where we used the Vpython module [Scherer et al., 2000] to generate the graphics. We have set the sampling rate to 200 Hz ($h = 5$ ms) for all the examples, unless explicitly mentioned otherwise. The control input value is updated at every other time step meaning that the update rate is 100 Hz. The simulation is stopped if the trajectory reaches to the boundary of D or a timeout is reached without satisfying the objective. Moreover, if the regulation objective is to reach a point other than the origin ($x \equiv 0$), we consider the cost (5.2), the value (5.8), and the obtained differential equation of the value parameters (5.13) by redefining $x := x - x_{\text{ref}}$.

5.5.1 Systems Identifiable in Terms of a Given Set of Bases

In the following two examples, we assume that the bases constituting the system dynamics exist in Φ. The system identified, after running the proposed learning algorithm and obtaining the value function, clearly depends on the identification algorithm used and its tuning parameters.

In Table 5.1, we illustrate the variations of the identified system and the corresponding value function by implementing the presented SOL algorithm with the exact $\dot{x}$ and with the first-order approximation of the derivative. It can be observed that, in the pendulum example, both of the obtained equations match the exact system (5.23) with a good precision. On the other hand, the Lorenz system is a more challenging system. Hence, by the first-order approximation of $\dot{x}$ with $h = 5$ ms, only an approximation of the dynamics can be obtained, while the exact system (5.24) is identified if we use the exact $\dot{x}$. As shown in Figures 5.1 and 5.2, although the model obtained for Lorenz system by using the approximate state variables does not closely match the exact dynamics, the obtained controller can successfully solve the regulation problem as long as the prediction errors remain bounded.

Table 5.1 The system dynamics and the corresponding value function obtained by the proposed method, where the exact and the approximated derivatives of the state variables are used in different scenarios

Obtained with exact $\dot{x}$	Obtained with $\dot{x} \approx (x_{k+1} - x_k)/h, h = 5\,\text{ms}$

$$\text{Pendulum } (\Phi = \{1, x, \sin x\})$$

Obtained with exact $\dot{x}$	Obtained with $\dot{x} \approx (x_{k+1} - x_k)/h$
$\dot{x}_1 = -1.000x_2$	$\dot{x}_1 = -1.011x_2$
$\dot{x}_2 = -1.000x_2 - 19.600\sin(x_1) + 40.000u$	$\dot{x}_2 = -0.995x_2 - 19.665\sin(x_1) + 40.098u$
$V(x) = 1.974x_1^2 - 0.058x_2x_1 + 0.036x_2^2$	$V(x) = 2.049x_1^2 - 0.058x_2x_1 + 0.036x_2^2$
$\quad -2.2\sin(x_1)x_1 - 0.077\sin(x_1)x_2$	$\quad -2.371\sin(x_1)x_1 - 0.077\sin(x_1)x_2$
$\quad +1.548\sin^2(x_1)$	$\quad +1.630\sin^2(x_1)$

$$\text{Chaotic Lorenz system } (\Phi = \{1, x, x^2, x^3, x_i x_j\}, \quad i, j \in \{1, \dots, n\}, i \neq j)$$

Obtained with exact $\dot{x}$	Obtained with $\dot{x} \approx (x_{k+1} - x_k)/h$
$\dot{x}_1 = -10.000x_1 + 10.000x_2 + 1.000u$	$\dot{x}_1 = -10.070x_1 + 9.973x_2 + 0.989u$
$\dot{x}_2 = 28.000x_1 - 1.000x_2 - 1.000x_1x_3$	$\dot{x}_2 = 0.993x_1 - 0.997x_2 + 8.483x_3 - 1.000x_1x_3$
$\dot{x}_3 = -2.667x_3 + 1.000x_1x_2$	$\dot{x}_3 = -8.483x_1 - 8.483x_2 - 2.666x_3 + 1.000x_1x_2$
	$V(x) = 11.193x_1^2 + 8.389x_2x_1 + 42.855x_2^2$
$V(x) = 30.377x_1^2 + 48.939x_2x_1 + 25.311x_2^2$	$\quad -20.950x_3x_1 + 28.441x_3x_2 + 32.045x_3^2$
$\quad +1.500x_3^2 - 1.873x_1x_2x_3 + 4.719x_1^2x_2^2$	$\quad -1.899x_1^2x_2 - 4.777x_1x_2^2 - 0.456x_1x_2x_3$
$\quad -3.291x_1^2x_3 + 1.469x_1x_3^2 - 0.012x_1^2x_2x_3$	$\quad +5.064x_1^2x_2^2 + 2.953x_1^2x_3 - 8.168x_1x_3^2$
	$\quad -2.633x_1^2x_3x_2 + 1.353x_1^2x_3^2$

Figure 5.1 Responses of the Lorenz system while learning by using the approximated state derivatives as in Table 5.1, where starting from one equilibrium point, we regulated the system to another unstable equilibrium.

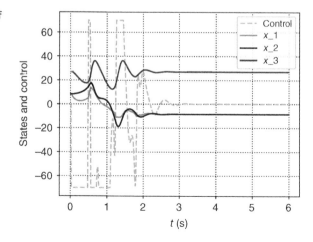

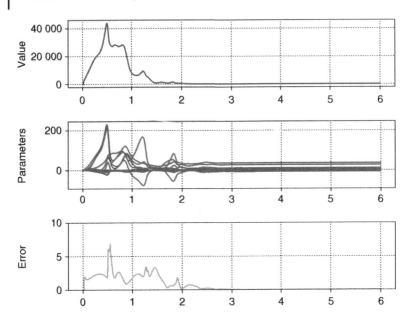

Figure 5.2 The value, components of P, and prediction error corresponding to Figure 5.1, respectively.

Example 1 (Pendulum)

The state space description of the system is given as follows:

$$\dot{x}_1 = -x_2,$$
$$\dot{x}_2 = -\frac{g}{l}\sin(x_1) - \frac{k}{m}x_2 + \frac{1}{ml^2}u, \qquad (5.23)$$

where $m = 0.1\,\text{kg}$, $l = 0.5\,\text{m}$, $k = 0.1$, and $g = 9.8\,\text{m/s}^2$. The performance criteria are defined by the choices of $Q = \text{diag}([1,1])$, $R = 2$.

Objective: The system is controlled to stabilize the otherwise unstable equilibrium point given by $x_{\text{ref}} \equiv 0$.

In Table 5.1, the learned dynamics and value function are listed for the exact and the approximated $\dot{x}$.

Example 2 (Chaotic Lorenz System)

The system dynamics are defined by

$$\dot{x}_1 = \sigma(x_2 - x_1) + u,$$
$$\dot{x}_2 = -x_2 + x_1(\rho - x_3),$$
$$\dot{x}_3 = x_1 x_2 - \beta x_3, \qquad (5.24)$$

where $\sigma = 10$, $\rho = 28$, and $\beta = 8/3$. Furthermore, we set the performance criteria to $Q = \mathrm{diag}([160,160,12])$, $R = 1$. This system has two unstable equilibrium points $(\pm\sqrt{72}, \pm\sqrt{72}, 27)$, where the trajectories of the system oscillate around these points.

Objective: By randomly setting the initial state

$$x_0 \in \{x \mid -40 \le x_i \le 40, i = 1, 2, 3\},$$

we stabilize the system around the unstable equilibrium $(-\sqrt{72}, -\sqrt{72}, 27)$.

5.5.2 Systems to Be Approximated by a Given Set of Bases

In what follows, we apply the presented learning scheme on another two benchmark examples. Unlike the previous examples, the dynamics of these systems include some rational terms that cannot be written in terms of some basis functions. However, an approximation can be obtained locally that is shown to be sufficient for successfully solving the regulation problem, as shown in Figures 5.3–5.10.

Moreover, as shown in Figure 5.11, a video of the graphical simulation of the following benchmark examples is included.

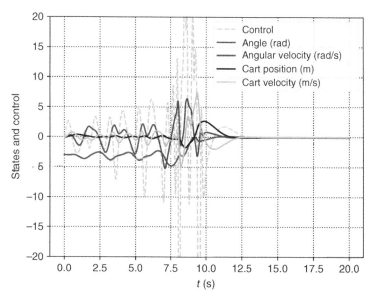

Figure 5.3 Responses of the cartpole system while learning by using the approximated state derivatives.

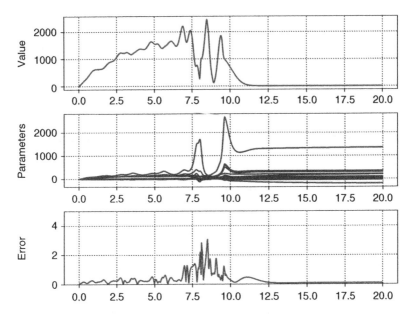

Figure 5.4 The value, components of P, and prediction error corresponding to Figure 5.3, respectively.

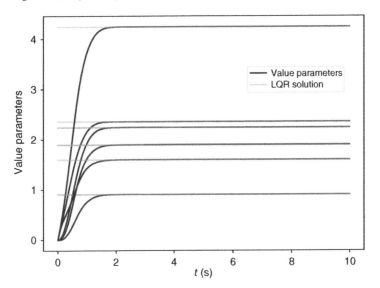

Figure 5.5 Convergence of the components of P to the LQR solution, obtained by solving the algebraic Riccati equation, is shown. Accordingly, it is evident that by running SOL algorithm on the linear system, the parameters of the value converge to the LQR solution.

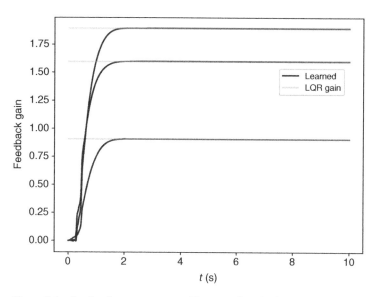

Figure 5.6 For the linear system, we illustrate that the feedback gain obtained by the learning algorithm presented asymptotically converges to the optimal gain given by the LQR solution in Figure 5.5.

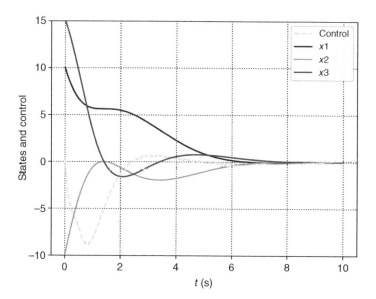

Figure 5.7 The state trajectories and the control signal while learning that correspond to Figures 5.5 and 5.6.

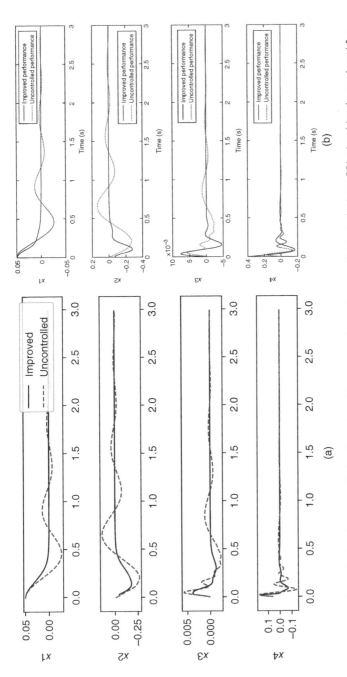

Figure 5.8 (a). Controlled and uncontrolled trajectories of the suspension system are illustrated where SOL algorithm is employed for learning the control. (b). Uncontrolled and controlled trajectories using the technique presented in Jiang and Jiang [2017]/John Wiley & Sons. By comparing the performances of two learned controllers, one can verify that the results are comparable. Accordingly, SOL can be alternatively used to approximate the optimal control for unknown systems.

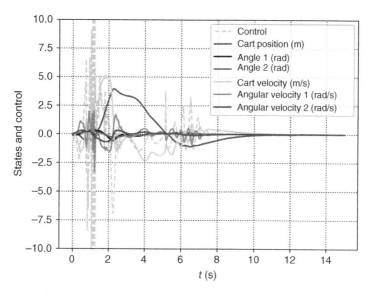

Figure 5.9 Responses of the double-inverted pendulum system while learning by using the approximated state derivatives.

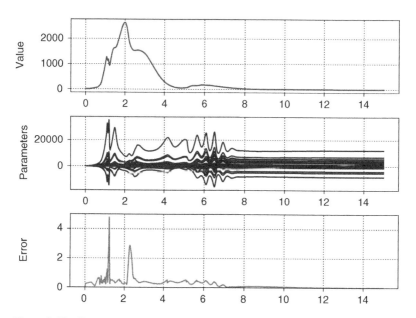

Figure 5.10 The value, components of P, and prediction error corresponding to Figure 5.9, respectively.

Figure 5.11 A view of the graphical simulations of the benchmark cartpole and double inverted pendulum examples. The video can be accessed in: https://youtu.be/-j0vaHE9MZY.

Example 3 (Cartpole Swing Up)

The dynamics are given as follows:

$$\dot{x}_1 = x_2,$$

$$\dot{x}_2 = \frac{-u\cos(x_1) - mLx_2^2\sin(x_1)\cos(x_1) + (M+m)g\sin(x_1)}{L(M + m\sin(x_1)^2)},$$

$$\dot{x}_3 = x_4,$$

$$\dot{x}_4 = \frac{u + m\sin(x_1)(Lx_2^2 - g\cos(x_1))}{M + m\sin(x_1)^2}, \tag{5.25}$$

where the state vector is composed of the angle of the pendulum from upright position, the angular velocity, and the position and velocity of the cart, with $m = 0.1\,\text{kg}$, $M = 1\,\text{kg}$, $L = 0.8\,\text{m}$, and $g = 9.8\,\text{m/s}^2$. Moreover, we choose $Q = \text{diag}([60, 1.5, 180, 45])$, $R = 1$.

Objective: By starting from some initial angles close to the stable angle of the pendulum $(\pm\pi)$, the cart swings up the pendulum to reach to and stay around the unstable equilibrium given as $x_{\text{ref}} \equiv 0$.

By running the learning scheme, an approximation of the system is identified as follows:

$$\dot{x}_1 = 1.000x_2,$$

$$\dot{x}_2 = 12.934\sin(x_1) + 0.230\sin(x_3) - 1.234\cos(x_1)u,$$

$$\dot{x}_3 = 0.995x_4,$$

$$\dot{x}_4 = 0.926\sin(x_1) + 0.953u, \tag{5.26}$$

where $\Phi = \{1, x, x^2, x^3, \sin x, \cos x\}$. Moreover, considering the assumed bases, we obtained the optimal value function as follows:

$$
\begin{aligned}
V(x) = {} & 59.712x_1^2 + 9.855x_2x_1 + 134.855x_2^2 + 9.587x_3x_1 \\
& + 241.295x_3x_2 + 223.389x_3^2 + 4.418x_4x_1 + 222.022x_4x_2 \\
& + 226.646x_4x_3 + 100.417x_4^2 - 63.050\sin(x_1)x_1 \\
& + 1098.765\sin(x_1)x_2 + 2294.259\sin^2(x_1)
\end{aligned}
$$

$$+ 984.786 \sin(x_1)x_3 + 909.030 \sin(x_1)x_4 - 1.712 \sin(x_3)x_1$$
$$+ 18.102 \sin(x_3)x_2 + 15.812 \sin(x_3)x_3 + 0.806 \sin^2(x_3)$$
$$+ 75.231 \sin(x_3)\sin(x_1) + 15.072 \sin(x_3)x_4. \tag{5.27}$$

Example 4 (Double-Inverted Pendulum on a Cart)
By defining $y := [q \quad \theta_1 \quad \theta_2]^T$ to be a vector of the cart position and angles of the double pendulum from the top equilibrium point, the system dynamics can be written in the following form:

$$\dot{x} = \begin{bmatrix} \dot{y} \\ M^{-1}f(y,\dot{y}) \end{bmatrix}, \tag{5.28}$$

where

$$M = \begin{bmatrix} m + m_1 + m_2 & l_1(m_1 + m_2)\cos(\theta_1) & m_2 l_2 \cos(\theta_2) \\ l_1(m_1 + m_2)\cos(\theta_1) & l_1^2(m_1 + m_2) & l_1 l_2 m_2 \cos(\theta_1 - \theta_2) \\ l_2 m_2 \cos(\theta_2) & l_1 l_2 m_2 \cos(\theta_1 - \theta_2) & l_2^2 m_2 \end{bmatrix},$$

$$f(y,\dot{y}) = \begin{bmatrix} l_1(m_1 + m_2)\dot{\theta}_1^2 \sin(\theta_1) + m_2 l_2 \dot{\theta}_2^2 \sin(\theta_2) - d_1 \dot{q} + u \\ -l_1 l_2 m_2 \dot{\theta}_2^2 \sin(\theta_1 - \theta_2) + g(m_1 + m_2)l_1 \sin(\theta_1) - d_2 \theta_1 \\ l_1 l_2 m_2 \dot{\theta}_1^2 \sin(\theta_1 - \theta_2) + g l_2 m_2 \sin(\theta_2) - d_3 \theta_2 \end{bmatrix},$$

$m = 6\,\text{kg}, m_1 = 3\,\text{kg}, m_2 = 1\,\text{kg}, l_1 = 1\,\text{m}, l_2 = 2\,\text{m}, d_1 = 10, d_2 = 1,$ and $d_3 = 0.5$.

Objective: We run the system from random angles around the top unstable equilibrium of the pendulums given by $\theta_1 = 0$ and $\theta_2 = 0$, where the controller has to learn to regulate the system to $x_{\text{ref}} \equiv 0$.

We choose the bases as $\Phi = \{1, x, x^2\}$. Moreover, the performance criteria are defined by $Q = \text{diag}([15, 15, 15, 1, 1, 1])$ and $R = 1$. A sample of the obtained approximate dynamics is

$$\dot{x}_1 = 0.998x_4, \quad \dot{x}_2 = 0.997x_5, \quad \dot{x}_3 = 0.996x_6,$$

$$\dot{x}_4 = 0.238x_1 - 4.569x_2 1.245x_3 - 1.891x_4 - 0.908x_6$$
$$- 0.105x_2^2 + 5.0131u - 2.824x_2^2 u,$$

$$\dot{x}_5 = 16.718x_2 - 2.328x_3 + 1.558x_4 - 0.598x_5 + 0.130x_6$$
$$- 0.114x_2^2 - 4.9911u 5.777x_2^2 u - 0.690x_3^2 u,$$

$$\dot{x}_6 = 0.123x_1 - 6.721x_2 + 9.032x_3 + 0.191x_5 - 0.358x_6$$
$$+ 0.969x_3^2 + 0.184x_6^2 - 1.898x_2^2 u + 1.431x_3^2 u. \tag{5.29}$$

It should be noted that because of the random initial conditions and different samples in the database, a different approximation of the system may be obtained in any learning procedure. Furthermore, considering the dimension of the system and the number of terms in the identified system (5.29), the obtained value function includes many terms of polynomials as expected. Therefore, for the sake of brevity, the obtained optimal value function is omitted.

5.5.3 Comparison Results

In this section, we compare the results obtained by SOL with other techniques in the literature. We present the comparison results on two examples. In the first example, we consider an unknown linear system. Then, we compare the value parameters obtained with the LQR solution of the same system. In the second example, we compare learning a nonlinear system with another learning technique presented in the literature.

Example 5 (A Linear System)
Consider the following linear system:

$$\dot{x} = \begin{bmatrix} 0 & 1 & 0 \\ 0 & 0 & 1 \\ -0.1 & -0.5 & -0.7 \end{bmatrix} x + \begin{bmatrix} 0 \\ 0 \\ 1 \end{bmatrix} u, \tag{5.30}$$

taken from Jiang and Jiang [2017]. The objective is defined by choosing $Q = I_3$ and $R = 1$.

Through this example, we investigate the convergence of SOL algorithm to the optimal control. Considering that the system is linear, one can compute the optimal solution by solving ARE. We compare the resulting solution with the value parameters P obtained by SOL, where the chosen vector of bases includes only the constant and linear terms, i.e. $\Phi = \begin{bmatrix} 1 & x^T \end{bmatrix}$. In Figure 5.5, it can be clearly observed that solutions of (5.13) converges to the optimal parameters obtained by solving ARE.

Furthermore, in Figure 5.6, we compare the feedback gain obtained by SOL with the optimal feedback gain. It can be verified that since the system is linear, the feedback control rule (5.11) converts to a linear feedback with a constant gain. In Figure 5.6, we compare this gain with the optimal gain obtained by solving the LQR problem, where the convergence to the optimal gains are evident.

Finally, in Figure 5.7, the state trajectories and the control signal demonstrate the asymptotic stability of the system under learning policy.

Example 6 (Car Suspension System)
In this example, we compare the presented learning algorithm with the RL technique employed in Jiang and Jiang [2017] on a nonlinear system. Consider the following nonlinear suspension system:

$$\dot{x}_1 = x_2,$$
$$x_2 = -\frac{k_s(x_1 - x_3) + k_n(x_1 - x_3)^3 + b_s(x_2 - x_4) - cu}{m_b},$$

$$\dot{x}_3 = x_3,$$

$$\dot{x}_4 = \frac{k_s(x_1 - x_3) + k_n(x_1 - x_3)^3 + b_s(x_2 - x_4) - k_t x_3 - cu}{m_w},$$

where the coefficients are given as $k_s = 16\,000$, $k_n = \frac{k_s}{10}$, $kt = 190\,000$, $c = 10\,000$, $m_b = 300$, $m_w = 60$, and $b_s = 1000$. The LQR objective is defined by $Q = \text{diag}([100,1,1,1])$, and $R = 1$.

To implement the SOL algorithm, we choose the polynomial bases of orders up to three. Moreover, we add a periodic exploration signal with a small amplitude to the learning control as $u_e = 0.01 \sin(50t)$ that allows convergence of the system identifier.

In Jiang and Jiang [2017], the authors employ a Sum of Squares (SOS) programming-based RL approach to learn the optimal control for this nonlinear system. To compare the MBRL technique presented with Jiang and Jiang [2017], we illustrate the trajectories of the system under both learned controllers in Figure 5.8, b with the same initial conditions. The uncontrolled suspension system is itself asymptotically stable. However, the convergence results can be improved by using a controller. According to Figure 5.8, b, it is evident that both of the techniques can efficiently learn the similar improved control policy.

5.6 Summary

Considering the control regulation problem, the structured dynamics helped us in analytically computing an iterative update rule to improve the optimal value function. In a model-based learning framework, we update the value according to the latest model identified. We applied the learning algorithm presented on different benchmark examples. Moreover, the comparison results demonstrated the convergence of the learned control to the optimal control. Based on the computational complexity and the performance observed in the numerical and graphical simulations, we showed some potential opportunities in employing the SOL algorithm as an online model-based learning technique.

Bibliography

S. N. Balakrishnan, Jie Ding, and Frank L. Lewis. Issues on stability of ADP feedback controllers for dynamical systems. *IEEE Transactions on Systems, Man, and Cybernetics, Part B (Cybernetics)*, 38(4):913–917, 2008.

Shubhendu Bhasin, Rushikesh Kamalapurkar, Marcus Johnson, Kyriakos G. Vamvoudakis, Frank L. Lewis, and Warren E. Dixon. A novel actor–critic–identifier architecture for approximate optimal control of uncertain nonlinear systems. *Automatica*, 49(1):82–92, 2013.

Steven L. Brunton, Joshua L. Proctor, and J. Nathan Kutz. Discovering governing equations from data by sparse identification of nonlinear dynamical systems. *Proceedings of the National Academy of Sciences of the United States of America*, 113(15):3932–3937, 2016.

Vladimir Gaitsgory, Lars Grüne, and Neil Thatcher. Stabilization with discounted optimal control. *Systems & Control Letters*, 82:91–98, 2015.

Yu Jiang and Zhong-Ping Jiang. *Robust Adaptive Dynamic Programming*. John Wiley & Sons, 2017.

Eurika Kaiser, J. Nathan Kutz, and Steven L. Brunton. Sparse identification of nonlinear dynamics for model predictive control in the low-data limit. *Proceedings of the Royal Society A*, 474(2219):20180335, 2018.

Rushikesh Kamalapurkar, Joel A. Rosenfeld, and Warren E. Dixon. Efficient model-based reinforcement learning for approximate online optimal control. *Automatica*, 74:247–258, 2016a.

Rushikesh Kamalapurkar, Patrick Walters, and Warren E. Dixon. Model-based reinforcement learning for approximate optimal regulation. *Automatica*, 64(C):94–104, 2016b.

Wei Kang and Lucas C. Wilcox. Mitigating the curse of dimensionality: Sparse grid characteristics method for optimal feedback control and hjb equations. *Computational Optimization and Applications*, 68(2):289–315, 2017.

Idris Kharroubi, Nicolas Langrené, and Huyên Pham. A numerical algorithm for fully nonlinear HJB equations: An approach by control randomization. *Monte Carlo Methods and Applications*, 20(2):145–165, 2014.

Jyrki Kivinen, Alexander J. Smola, and Robert C. Williamson. Online learning with kernels. *IEEE Transactions on Signal Processing*, 52(8):2165–2176, 2004.

Frank L. Lewis and Derong Liu. *Reinforcement Learning and Approximate Dynamic Programming for Feedback Control*. John Wiley & Sons, 2013.

Frank L. Lewis and Draguna Vrabie. Reinforcement learning and adaptive dynamic programming for feedback control. *IEEE Circuits and Systems Magazine*, 9(3):32–50, 2009.

William M. McEneaney. A curse-of-dimensionality-free numerical method for solution of certain HJB PDEs. *SIAM Journal on Control and Optimization*, 46(4):1239–1276, 2007.

Romain Postoyan, L. Buşoniu, D. Nešić, and Jamal Daafouz. Stability of infinite-horizon optimal control with discounted cost. In *53rd IEEE Conference on Decision and Control*, pages 3903–3908. IEEE, 2014.

Anna Prach, Ozan Tekinalp, and Dennis S. Bernstein. Infinite-horizon linear-quadratic control by forward propagation of the differential Riccati equation. *IEEE Control Systems Magazine*, 35(2):78–93, 2015.

Zhihua Qu and Jian-Xin Xu. Model-based learning controls and their comparisons using lyapunov direct method. *Asian Journal of Control*, 4(1):99–110, 2002.

David Scherer, Paul Dubois, and Bruce Sherwood. VPython: 3D interactive scientific graphics for students. *Computing in Science & Engineering*, 2(5):56–62, 2000.

Steven Van Vaerenbergh and Ignacio Santamaría. Online regression with kernels. In *Regularization, Optimization, Kernels, and Support Vector Machines*, Johan A.K. Suykens, Marco Signoretto, and Andreas Argyriou (eds.), pages 495–521. CRC Press, 2014.

Fei-Yue Wang, Huaguang Zhang, and Derong Liu. Adaptive dynamic programming: An introduction. *IEEE Computational Intelligence Magazine*, 4(2):39–47, 2009.

Avishai Weiss, Ilya Kolmanovsky, and Dennis S. Bernstein. Forward-integration Riccati-based output-feedback control of linear time-varying systems. In *Proceedings of the American Control Conference*, pages 6708–6714. IEEE, 2012.

Bin Xian, Darren M. Dawson, Marcio S. de Queiroz, and Jian Chen. A continuous asymptotic tracking control strategy for uncertain nonlinear systems. *IEEE Transactions on Automatic Control*, 49(7):1206–1211, 2004.

Huaguang Zhang, Lili Cui, Xin Zhang, and Yanhong Luo. Data driven robust approximate optimal tracking control for unknown general nonlinear systems using adaptive dynamic programming method. *IEEE Transactions on Neural Networks*, 22(12):2226–2236, 2011.

6

A Structured Online Learning Approach to Nonlinear Tracking with Unknown Dynamics

One of the most common problems in the control of dynamical systems is to track a desired reference trajectory, which is found in a variety of real-world applications. In this chapter, we extend the Structured Online Learning (SOL) framework to tracking with unknown dynamics. Similar to regulation problems, the applications of tracking control can benefit from Model-based Reinforcement Learning (MBRL) that can handle the parameter updates more efficiently. The results presented in this chapter are published in Farsi and Liu [2021].

6.1 Introduction

Tracking a desired reference trajectory is one of the most classical objectives in the control of dynamical systems and is commonly encountered in many real-world applications. However, the design of an effective tracking controller via conventional approaches often requires sufficient knowledge of the model, and a lot of calculations and considerations are involved for any particular application. On the other hand, Reinforcement Learning (RL) techniques suggest a more adaptable framework that requires less knowledge about the system dynamics.

In this chapter, we extend the results obtained in Chapter 5 to the tracking problem of unknown continuous dynamical systems. In Section 6.2, we propose an approximate optimal tracking control framework based on a particular structure of nonlinear dynamics, where a linear quadratic discounted cost is assumed. Section 6.3 provides the details of implementation of the obtained framework as a learning-based approach. In Section 6.4, two numerical results illustrating the proposed approach are reported on two benchmark examples.

Model-Based Reinforcement Learning: From Data to Continuous Actions with a Python-based Toolbox, First Edition. Milad Farsi and Jun Liu.

6.2 A Structured Online Learning for Tracking Control

Consider the nonlinear control-affine system

$$\dot{x} = f(x) + g(x)u, \tag{6.1}$$

where $x \in D \subseteq \mathbb{R}^n$ and $u \in U \subseteq \mathbb{R}^m$, are the state and the control input, respectively. Moreover, $f : D \to \mathbb{R}^n$ and $g : D \to \mathbb{R}^{n \times m}$.

Assumption 6.1 *Each component of f and g can be identified or effectively approximated within the compact domain of interest by a linear combination of some bases functions $\phi_i \in C^1 : D \to \mathbb{R}$ for $i = 1, 2, \ldots, p$.*

Accordingly, (6.1) is rewritten as follows:

$$\dot{x} = W\Phi(x) + \sum_{j=1}^{m} W_j \Phi(x) u_j, \tag{6.2}$$

where W and $W_j \in \mathbb{R}^{n \times p}$ are the matrices of the coefficients obtained for $j = 1, 2, \ldots, m$, and

$$\Phi(x) = [x^T \ \phi_{n+1}(x) \ \ldots \ \phi_p(x)]^T.$$

Remark 6.1 The structure employed in (6.2) is motivated by the fact that the linear combination of the bases provides an opportunity of obtaining an analytical control approach. At the same time, we can profit from the variety of the applicable identification techniques.

Remark 6.2 Regarding Assumption 6.1, in this chapter, we will not perform convergence analysis of any particular identification approach. Instead, we will focus on the controller design procedure, in a way that the problem formulation and the design allow exploiting different identification methods alternatively. Hence, this section presents the control technique for the given W and W_j. However, in the implementations, an estimation of these weights will be used, which is discussed in Section 6.3.

Assumption 6.2 *The given reference trajectory $y_{\text{ref}}(t) : \mathbb{R} \to \mathbb{R}^d$ is a particular solution of a dynamical system of the form*

$$\dot{y}_{\text{ref}} = M\Psi(y_{\text{ref}}), \tag{6.3}$$

where

$$\Psi(y_{\text{ref}}) = [y_{\text{ref}}^T \ \psi_{d+1}(y_{\text{ref}}) \ \ldots \ \psi_q(y_{\text{ref}})]^T$$

is a set of bases and $M \in \mathbb{R}^{d \times q}$ is the matrix of coefficients. This system can be seen as the virtual command generator, which with the variety of the chosen bases can *support a wide range of signals, from simple ramp or sinusoidal signals to more complex ones.*

In the optimal tracking problem, a cost functional is assumed to measure the performance. Starting from the initial condition $x_0 = x(0)$, the following discounted linear quadratic cost is minimized along the trajectory:

$$J(x_0, u)$$
$$= \lim_{T \to \infty} \int_0^T e^{-\gamma t} \left((Cx - y_{ref})^T Q(Cx - y_{ref}) + u^T Ru \right) dt, \tag{6.4}$$

where $Q \in \mathbb{R}^{d \times d}$ is positive semidefinite, $\gamma \geq 0$ is the discount factor, and $R \in \mathbb{R}^{m \times m}$ is a diagonal matrix with only positive values, given by design criteria. Moreover, corresponding to the dimension of y_{ref}, a subset of the states is chosen by using $C \in \mathbb{R}^{d \times n}$, that includes entries one corresponding to the measured states and zero elsewhere.

For the closed-loop system, by assuming a feedback control law

$$u = \omega(x(t), y_{ref}(t))$$

for $t \in [0, \infty)$, the optimal control is given by

$$\omega^* = \underset{u(\cdot) \in \Gamma(x_0)}{\text{argmin}} \ J(x_0, u(\cdot)), \tag{6.5}$$

where $\Gamma(x_0)$ is the set of admissible control signals.

Lemma 6.1 *The optimal tracking control obtained by minimizing*

$$J(x_0, u) = \lim_{T \to \infty} \int_0^T e^{-\gamma t} \left(\bar{\Phi}^T \bar{Q} \bar{\Phi} + u^T Ru \right) dt, \tag{6.6}$$

is equivalent to the solution of (6.5) assuming (6.4), where

$$\bar{\Phi}(x, y_{ref}) = \left[(Cx)^T \ y_{ref}^T \ \phi_{(d+1)}(x) \ \dots \ \phi_{(p)}(x) \right.$$
$$\left. \psi_{(d+1)}(y_{ref}) \ \dots \ \psi_{(q)}(y_{ref}) \right]^T, \tag{6.7}$$

and

$$\bar{Q} = \text{diag} \left(\begin{bmatrix} Q & -Q \\ -Q & Q \end{bmatrix}, [\mathbf{0}_{(p+q-2d) \times (p+q-2d)}] \right) \tag{6.8}$$

is a block diagonal matrix that contains all zeros except the first block which correspond to the linear bases Cx and y_{ref}.

Proof: It is straightforward to rewrite the performance measure (6.4) in terms of the vector of bases $[(Cx)^T \ y_{ref}^T]$ with a positive semidefinite matrix defined as the none-zero block in (6.8). Later, the obtained cost is again transformed to the space of bases $\bar{\Phi}$ to take the form (6.6), where we assume (6.8). ∎

Remark 6.3 In some particular applications, while tracking a given trajectory, we might need to penalize at the same time the growth in some other states which are not in the list of the tracked states y_{ref}. Such conditions can be still handled by the block-diagonal matrix (6.8), that is to assign a nonzero value, corresponding to that particular state, in the diagonal of the second block of (6.8).

Now, consider the system dynamics (6.2) and the command generator (6.3). We define the augmented system as follows:

$$\dot{z} = \begin{bmatrix} \dot{x} \\ \dot{y}_{\text{ref}} \end{bmatrix} = F\bar{\Phi} + \sum_{j=1}^{m} G_j \bar{\Phi} u_j, \tag{6.9}$$

where the system matrices F and G_j are obtained by rearranging the block entries of the coefficient matrices of (6.2) and (6.3) according to the ordering of entries in $\bar{\Phi}$.

By defining the Hamiltonian, the corresponding Hamilton–Jacobi–Bellman (HJB) equation of (6.6) is given as follows:

$$-\frac{\partial}{\partial t}(e^{-\gamma t}V) = \min_{u(\cdot)\in\Gamma(x_0)} \left\{ H = e^{-\gamma t}\left(\bar{\Phi}^T\bar{Q}\bar{\Phi} + u^T R u\right) \right.$$

$$\left. +e^{-\gamma t}\frac{\partial V}{\partial z}\left(F\bar{\Phi} + \sum_{j=1}^{m} G_j\bar{\Phi}u_j\right)\right\}. \tag{6.10}$$

Then, we employ an approximation scheme to estimate a parameterized value function $V : D \times [0, \infty) \to \mathbb{R}$ satisfying the above partial differential equation.

Unlike other approximate optimal control approaches in the literature, such as Zhang et al. [2011], Bhasin et al. [2013], Kamalapurkar et al. [2016a], and Zhu et al. [2016], we use a quadratic form to parameterize the value function as follows:

$$V = \bar{\Phi}^T P\bar{\Phi}, \tag{6.11}$$

where P is a symmetric matrix.

As suggested in Farsi and Liu [2020] and discussed in Chapter 5, defining V in the product space $\Lambda := \bar{\Phi} \times \bar{\Phi}$ provides a compact form of parameterizing the value function, and a variety of bases can be produced in the product space Λ by only including a limited number of bases in $\bar{\Phi}$. It should be noted that updating the parameters in the matrix form in SOL will considerably decrease the computations required to update the parameters, where the matrix multiplications involved will remain as cheap as the dimension of $\bar{\Phi}$. On the other hand, in alternative parameter update methods, such as gradient descent, matrix multiplications of the dimension of the numbers of elements in set Λ is involved, which is a considerably larger set compared to $\bar{\Phi}$.

Assuming (6.11), the Hamiltonian is written as follows:

$$
H = e^{-\gamma t}(\bar{\Phi}^T \bar{Q} \bar{\Phi} + u^T R u)
$$

$$
+ e^{-\gamma t} \left(\frac{\partial \bar{\Phi}}{\partial z}^T P \bar{\Phi} \right)^T \left(F \bar{\Phi} + \sum_{j=1}^{m} G_j \bar{\Phi} u_j \right)
$$

$$
+ e^{-\gamma t} \left(F \bar{\Phi} + \sum_{j=1}^{m} G_j \Phi u_j \right)^T \left(\frac{\partial \bar{\Phi}}{\partial z}^T P \bar{\Phi} \right).
$$

In the following, we rewrite the quadratic term of u based on its components, assuming that $r_j \neq 0$ is the jth component on the diagonal of R:

$$
H = e^{-\gamma t} \left(\bar{\Phi}^T \bar{Q} \bar{\Phi} + \sum_{j=1}^{m} r_j u_j^2 + \bar{\Phi}^T P \frac{\partial \bar{\Phi}}{\partial z} F \bar{\Phi} \right.
$$

$$
+ \bar{\Phi}^T P \frac{\partial \bar{\Phi}}{\partial z} \left(\sum_{j=1}^{m} G_j \bar{\Phi} u_j \right) + \bar{\Phi}^T F^T \frac{\partial \bar{\Phi}}{\partial z}^T P \bar{\Phi}
$$

$$
\left. + \left(\sum_{j=1}^{m} u_j \bar{\Phi}^T G_j^T \right) \frac{\partial \bar{\Phi}}{\partial z}^T P \bar{\Phi} \right). \tag{6.12}
$$

The obtained Hamiltonian is minimized if, for the jth system input, we have

$$
\frac{\partial H}{\partial u_j} = 2r_j u_j + 2\bar{\Phi}^T P \frac{\partial \bar{\Phi}}{\partial z} G_j \bar{\Phi} \tag{6.13}
$$

$$
= 0, \qquad j = 1, 2, \dots, m.
$$

Accordingly, the optimal control input is calculated as follows:

$$
u_j^* = -\bar{\Phi}^T r_j^{-1} P \frac{\partial \bar{\Phi}}{\partial z} G_j \bar{\Phi}. \tag{6.14}
$$

Then, we substitute the obtained optimal feedback control law and the value function to (6.10), which yields

$$
- e^{-\gamma t} \bar{\Phi}^T \dot{P} \bar{\Phi} + \gamma e^{-\gamma t} \bar{\Phi}^T P \bar{\Phi}
$$

$$
= e^{-\gamma t} \left(\bar{\Phi}^T \bar{Q} \bar{\Phi} + + \bar{\Phi}^T P \frac{\partial \bar{\Phi}}{\partial z} F \bar{\Phi} + \bar{\Phi}^T F^T \frac{\partial \bar{\Phi}}{\partial z}^T P \bar{\Phi} \right.
$$

$$
+ \bar{\Phi}^T P \frac{\partial \bar{\Phi}}{\partial z} \left(\sum_{j=1}^{m} G_j \bar{\Phi} r_j^{-1} \bar{\Phi}^T G_j^T \right) \frac{\partial \bar{\Phi}}{\partial z}^T P \bar{\Phi}
$$

$$
\left. - 2 \bar{\Phi}^T P \frac{\partial \bar{\Phi}}{\partial z} \left(\sum_{j=1}^{m} G_j \bar{\Phi} r_j^{-1} \bar{\Phi}^T G_j^T \right) \frac{\partial \bar{\Phi}}{\partial z}^T P \bar{\Phi} \right).
$$

By some manipulation we get

$$-\Phi^T \dot{P} \bar{\Phi} + \gamma \bar{\Phi}^T P \bar{\Phi} = \bar{\Phi}^T \bar{Q} \bar{\Phi}$$

$$- \bar{\Phi}^T P \frac{\partial \bar{\Phi}}{\partial z} \left(\sum_{j=1}^{m} G_j \bar{\Phi} r_j^{-1} \bar{\Phi}^T G_j^T \right) \frac{\partial \bar{\Phi}}{\partial z}^T P \bar{\Phi}$$

$$+ \bar{\Phi}^T P \frac{\partial \bar{\Phi}}{\partial z} F \bar{\Phi} + \bar{\Phi}^T F^T \frac{\partial \bar{\Phi}}{\partial z}^T P \bar{\Phi}.$$

Finally, a sufficient condition for satisfying this equation is obtained as follows:

$$-\dot{P} = \bar{Q} + P \frac{\partial \bar{\Phi}}{\partial z} F + F^T \frac{\partial \bar{\Phi}}{\partial z}^T P - \gamma P$$

$$- P \frac{\partial \bar{\Phi}}{\partial z} \left(\sum_{j=1}^{m} G_j \bar{\Phi} r_j^{-1} \bar{\Phi}^T G_j^T \right) \frac{\partial \bar{\Phi}}{\partial z}^T P. \tag{6.15}$$

The standard way for solving such an equation is by integrating in the backward direction, where it requires full knowledge of the system including the weights F and G_j for all time horizons. As a result, a value of P is obtained, which realizes the optimal value function (6.11), and leads to the optimal control (6.14). Recalling Remark 6.2, one can employ the presented method to design a tracking control of a known system. However, in this chapter, we focus on the learning problem where the accurate system model may not be known at first place. Therefore, we propagate the obtained differential equation in the forward direction. This will provide an opportunity to update our estimation of the system dynamics online at any step together with the control rule.

6.2.1 Stability and Optimality in the Linear Case

In this section, the proposed control framework is analyzed in a special case where both the virtual target and the tracker systems are linear.

The optimality and stability of optimal control approaches based on forward propagation of the Riccati differential equation were discussed in Chapter 5 in detail. Accordingly, using Lemma 5.2, we can guarantee obtaining a stabilizing controller by employing the forward solution of differential Riccati equation, where in addition, the convergence to the Linear Quadratic Regulator (LQR) control is expected as $t \to \infty$.

Therefore, in the followings, by taking linear systems as an example and by making a connection to the classical LQR formulation, we will demonstrate that for the linear systems as the special case, the presented approach becomes equivalent to the LQR framework. This allows us to provide guarantees for the presented tracking control approach for the linear case.

Proposition 6.1 *Consider the controllable linear system $\dot{x} = Ax + Bu_1$ together with the linear command generator $\dot{y}_{ref} = Ay_{ref} + Bu_d$, where u_d is the given desired input. Furthermore, the tracking performance measure is given by (6.4) with $C = I_{n \times n}$. Then, as $\gamma \to 0$, the optimal feedback control (6.14) constructed by the solution of (6.15), and the optimal value (6.11) with the choice of $\bar{\Phi} = [x \quad y_{ref} \quad 1]^T$, approaches the LQR feedback control with the gain $k = r_1^{-1}\tilde{S}\tilde{B}$ of the generalized error system:*

$$\frac{d}{dt}\begin{bmatrix} x \\ e \\ 1 \end{bmatrix} = \tilde{A}\begin{bmatrix} x \\ e \\ 1 \end{bmatrix} + \tilde{B}u_1, \tag{6.16}$$

where $\tilde{A} = \begin{bmatrix} A & 0 & 0 \\ 0 & A & Bu_d \\ 0 & 0 & 0 \end{bmatrix}$, $\tilde{B} = \begin{bmatrix} B & -B & 0 \end{bmatrix}^T$, e is the tacking error, and S is given at any $t \in [0, \infty)$ by the solution of the well-known continues-time differential Riccati equation:

$$\dot{S} = \tilde{Q} - S\tilde{B}r_1^{-1}\tilde{B}^T S + S\tilde{A} + \tilde{A}^T S. \tag{6.17}$$

Proof: In this case, the linear bases of each tracker and the target dynamics, together with a constant basis 1 will suffice to implement the presented approach. Hence, we proceed with the choice of $\bar{\Phi} = [x \quad y_{ref} \quad 1]^T$. Note that this will not affect the generality and including extra basis will only add more columns of zeros in the computations. Then the augmented system (6.9) becomes

$$\dot{z} = \begin{bmatrix} \dot{x} \\ \dot{y}_{ref} \end{bmatrix} = \underbrace{\begin{bmatrix} A & 0 & 0 \\ 0 & A & Bu_d \end{bmatrix}}_{F}\bar{\Phi} + \underbrace{\begin{bmatrix} 0 & 0 & B \\ 0 & 0 & 0 \end{bmatrix}}_{G_1}\bar{\Phi}u_1. \tag{6.18}$$

Similar to the procedure presented in Section 6.2, the optimal control should satisfy the HJB equation

$$\bar{\Phi}^T P\bar{\Phi} + \gamma\bar{\Phi}^T P\bar{\Phi} = \bar{\Phi}^T\begin{bmatrix} Q & -Q & 0 \\ -Q & Q & 0 \\ 0 & 0 & 0 \end{bmatrix}\bar{\Phi} + u_1^{*T}r_1u_1^*$$

$$+ \left(\bar{\Phi}^T P\frac{\partial\bar{\Phi}}{\partial z}\right)\dot{z} + \dot{z}^T\left(\frac{\partial\bar{\Phi}}{\partial z}^T P\bar{\Phi}\right), \tag{6.19}$$

where

$$P = \begin{bmatrix} P_1 & P_2 & P_3 \\ P_2^T & P_4 & P_5 \\ P_3^T & P_5^T & P_6 \end{bmatrix}.$$

In this step, we substitute $y_{ref} = x + e$ in (6.19). Accordingly, we redefine $z := \begin{bmatrix} x & e \end{bmatrix}^T$ and $\bar{\Phi} := \begin{bmatrix} x & e & 1 \end{bmatrix}$. Hence, the augmented system (6.18) can be rewritten with the new definitions, where augmented system matrix become

$$G_1 := \begin{bmatrix} 0 & 0 & B \\ 0 & 0 & -B \end{bmatrix},$$

and F remain the same.

Now, by substituting the optimal control (6.14) and considering the change of variables, (6.19) is equivalent to

$$\bar{\Phi}^T \dot{S} \bar{\Phi} + \gamma \bar{\Phi}^T S \bar{\Phi} = \bar{\Phi}^T \tilde{Q} \bar{\Phi}$$
$$- \bar{\Phi}^T S \frac{\partial \bar{\Phi}}{\partial z} G_1 \bar{\Phi} r_1^{-1} \bar{\Phi}^T G_1^T \frac{\partial \bar{\Phi}}{\partial z}^T S \bar{\Phi}$$
$$+ \bar{\Phi}^T S \frac{\partial \bar{\Phi}}{\partial z} F \bar{\Phi} + \bar{\Phi}^T F^T \frac{\partial \bar{\Phi}}{\partial z}^T S \bar{\Phi}, \qquad (6.20)$$

where $\tilde{Q} = \text{diag}([0, Q, 0])$. For brevity, we omitted the detailed computations, while one can check with some effort the equivalency by using Lemma 6.1, and

$$S = \begin{bmatrix} P_1 + P_4 + P_2 + P_2^T & P_2 + P_4 & P_3 + P_5 \\ P_2^T + P_4^T & P_4 & P_5 \\ P_3^T + P_5^T & P_5^T & P_6 \end{bmatrix}.$$

By plugging in

$$\frac{\partial \bar{\Phi}}{\partial z} = \begin{bmatrix} I & 0 \\ 0 & I \\ 0 & 0 \end{bmatrix},$$

and defining

$$\tilde{A} = \frac{\partial \bar{\Phi}}{\partial z} F = \begin{bmatrix} A & 0 & 0 \\ 0 & A & B u_d \\ 0 & 0 & 0 \end{bmatrix},$$

$$\tilde{B} = \frac{\partial \bar{\Phi}}{\partial z} G_1 \bar{\Phi} = \begin{bmatrix} B & -B & 0 \end{bmatrix}^T,$$

we can conclude from (6.20):

$$\dot{S} = \tilde{Q} - S \tilde{B} r_1^{-1} \tilde{B}^T S + S \tilde{A} + \tilde{A}^T S - \gamma S. \qquad (6.21)$$

As $\gamma \to 0$, this yields the well-known continuous-time differential Riccati equation (6.17) for the generalized error system (6.16). Moreover, the LQR feedback gain $k = r_1^{-1} S \tilde{B}$ can be obtained by using the definition of $\tilde{B}$. ∎

For linear time-invariant systems, the steady-state solution of (6.17) realizes the optimal feedback control that guarantees asymptotic stability of the error and optimality of the solution. However, in the RL setting, as well as in time-variant systems, we can only rely on the present knowledge of the system and hence on the continuous evolutions of the differential Riccati equation. In the linear case of this type, the stability of the closed-loop system is guaranteed (see Lemma 5.2 where convergence analysis is provided by extending the results from Prach et al. [2015]).

Section 6.3 will discuss the details of implementing the nonlinear optimal tracking control as a MBRL approach.

6.3 Learning-based Tracking Control Using SOL

Initially, the SOL approach was proposed for solving stabilization and regulation problems [Farsi and Liu, 2020]. We discussed the SOL approach for regulation problems in detail in Chapter 5. In Section 6.2 of this chapter, we demonstrated that a tracking controller can be obtained by extending the underlying idea of SOL, which is somehow unifying the dynamics and the optimal control objective to gain some flexibility in solving the optimal control problem. As a result, we obtained a state-dependent matrix differential equation (6.15), whose solution provides the parameters of the nonlinear optimal tracking control in terms of the reference and state trajectories. In the followings, we will briefly review the model-based learning framework.

We run the system from some $x_0 \in D$ for a time step of length h. Then, by sampling the input, system state, and reference trajectories, we evaluate $\bar{\Phi}$, which together with the approximation of the augmented state derivative $\dot{z}$ is used to update our estimation of the augmented system, including the system and the commander dynamics coefficients. Later, the estimated weights and the current measured state is used to integrate (6.15) for some time step. This is immediately followed by updating the control value for the next iteration in the control loop by employing (6.14).

Consider the system weight update as follows:

$$[\hat{F} \; \hat{G}_1 \; \dots \; \hat{G}_j]_k = \underset{[F,G_1,G_j]_k}{\operatorname{argmin}} \; E(\dot{z}_k, [F, G_1, \dots G_j]_k, \Theta(z_k, u_k)),$$

where k is the time step and $E(\cdot)$ is the defined cost in the identification technique employed. Moreover, we construct

$$\Theta(z_k, u_k) = [\bar{\Phi}^T(z_k) \; \bar{\Phi}^T(z_k)u_{1k} \; \dots \; \bar{\Phi}^T(z_k)u_{mk}]^T \tag{6.22}$$

by using the measurements obtained from the system and reference trajectories at time t_k. To solve (6.22), different approaches can be alternatively exploited in various applications, such as Sparse Identification of Nonlinear Dynamics (SINDy) [Brunton et al., 2016], recursive least-squares, neural networks, each of which has its own benefits and drawbacks.

Remark 6.4   It should be noted that estimating the true dynamics by (6.22) is not generally a trivial task. However, this is a well-studied topic in the literature of system identification for any estimation technique specifically. The well-known persistent excitation of the input signal is one of the common requirements that can be satisfied by adding an exploring signal to the control.

Remark 6.5   Depending on the application, the commander dynamics (6.3) may be given or unknown. In the case the dynamics are provided, we initialize the associated coefficients when considering the augmented system. This will accelerate the identification process.

In the control update procedure, a recommended choice for initial condition P_0 is a zero square matrix of appropriate dimension. Then, off-the-shelf solvers can be used to effectively integrate (6.15). This also requires evaluations of $\partial \bar{\Phi}/\partial z_k$ at any time step. Since the bases $\bar{\Phi}$ are chosen beforehand, the partial derivatives can be analytically calculated and stored as functions. Hence, they can be evaluated for any z_k in a similar way as $\bar{\Phi}$ itself.

6.4   Simulation Results

To illustrate the effectiveness of the proposed SOL approach for tracking, we implement it on two benchmark nonlinear systems. As shown in Farsi and Liu [2020] and Chapter 5, SOL can be employed to solve the regulation problem of nonlinear systems with unknown dynamics, including the benchmark pendulum and Lorenz system examples. In what follows, we borrow these two examples to investigate the characteristics of the tracking control approach derived based on the SOL framework. For the simulation results, we use a Runge–Kutta solver to integrate the dynamics of the system. These data are treated as measurements in lieu of physical experiments.

In these benchmark examples, we update the model with the most recent measurement by using SINDy [Brunton et al., 2016]. However, considering that this identification technique is already studied and introduced as a data-efficient and robust method [Brunton et al., 2016; Kaiser et al., 2018], we will focus on the properties of the proposed control scheme rather than the identification process within the simulations.

Moreover, it is observed that the accuracy of the obtained model depends directly on the precision in measuring the derivatives of the states, which may be a source of noise in the real-world implementations, resulting in an ineffective controller. Through these simulations, we assume full access to the states. We then obtained the state derivatives using a one-step backward approximation, which can potentially be improved by considering more steps.

We performed the simulations in Python, including the 3D graphics generated via the Vpython module [Scherer et al., 2000]. The sampling rate is 200 Hz ($h = 5$ ms) for all the simulations. Accordingly, the control input value is calculated with the frequency of 100 Hz. In the learning process, to simulate the real behaviors of the given system, we integrate the differential equations with a sufficient precision from initial conditions randomly chosen within the domain of interest. However, we only allow measurements at time steps conforming to the sampling rate.

Furthermore, in the following examples, we use the candidate bases

$$\left\{1, x, x^2, x^3, \sin x, \cos x, x_i x_j\right\},$$

where $i, j \in \{1, \ldots, n\}, i \neq j$, and the operations on vector x is assumed to be componentwise and defines a subcategory of bases, e.g. $x^2 = \{x_1^2, \ldots, x_n^2\}$. Hence, Assumption 6.1 holds.

6.4.1 Tracking Control of the Pendulum

The state space model used for the simulation of the pendulum system is given by

$$\dot{x}_1 = -x_2,$$

$$\dot{x}_2 = -\frac{g}{l}\sin(x_1) - \frac{k}{m}x_2 + \frac{1}{ml^2}u, \tag{6.23}$$

where $m = 0.1$ kg, $l = 0.5$ m, $k = 0.1$, and $g = 9.8$ m/s^2. A desired performance is characterized by matrices $Q = \text{diag}([2, 7])$, $R = 1$, $\gamma = 1$.

The simulations for this system are done in two different scenarios. In the first scenario, we assume having the full state reference trajectory for the angle and angular velocity. Thus, we try, for instance, sinusoidal and ramp reference signals as below, respectively.

$$(a) : \begin{cases} y_{1_{\text{ref}}} = -\sin(t), \\ y_{2_{\text{ref}}} = \cos(t). \end{cases} \quad , \quad (b) : \begin{cases} y_{1_{\text{ref}}} = -t, \\ y_{2_{\text{ref}}} = 1, \end{cases} \tag{6.24}$$

where the corresponding states are measured by choosing $C = \text{diag}([1, 1])$ in (6.1).

Figures 6.1 and 6.2 illustrate system responses for sinusoidal and ramp references starting from a random initial condition, respectively. As seen in these figures, although the learning process is run with zero prior experience, it can efficiently track the reference by quickly learning the dynamics and the optimal tracking controller.

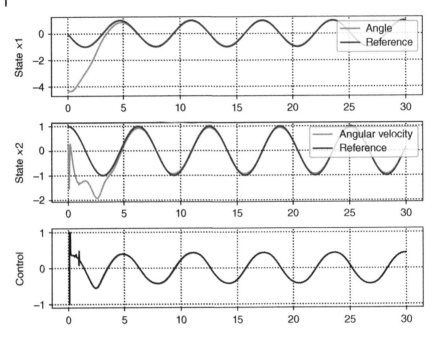

Figure 6.1 The control and states of the pendulum system within a run of the implemented learning approach from a randomly chosen initial condition for tracking a full-state sinusoidal reference signal provided as in (6.24a).

In the second scenario, to better examine the tracking control scheme presented in the main results, we assume only the trajectory of the angular position that is provided as the reference, where $C = \mathrm{diag}([1, 0])$ and $Q = 2$. Hence, the control objective is defined based on only the angular position error. For this reason, one should not expect the tracking results to be as smooth as the previous case. However, as shown in Figure 6.3, the objective is achieved by perfect tracking of the angular position, where, in addition, the other state still resembles the nongiven target trajectory, to an acceptable extent.

6.4.2 Synchronization of Chaotic Lorenz System

The Lorenz system is well known for its chaotic behavior around its unstable equilibrium points. As an illustrative example (Figure 6.4), we aim to synchronize two Lorenz systems starting from different initial conditions. This is done by measuring the states of one as the target system, then controlling the other system to track these target states over time by the proposed learning-based tracking control technique. We assume no prior knowledge of the system dynamics and parameters. Hence, the dynamics of the target and the tracker are learned together with

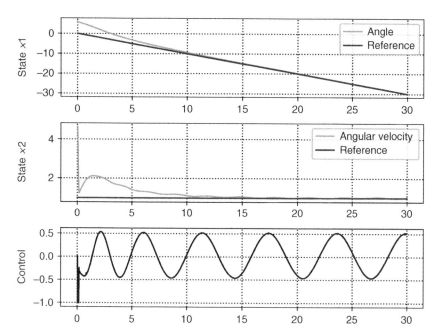

Figure 6.2 The control and states of the pendulum system within a run of the implemented learning approach from a randomly chosen initial condition for tracking a full-state ramp reference signal provided as in (6.24b).

a tracking controller on the fly. The system dynamics used in the simulation are described as follows:

$$\dot{x}_1 = \sigma(x_2 - x_1) + u,$$

$$\dot{x}_2 = -x_2 + x_1(\rho - x_3),$$

$$\dot{x}_3 = x_1 x_2 - \beta x_3, \tag{6.25}$$

where $\sigma = 10$, $\rho = 28$, and $\beta = 8/3$. Furthermore, we set the performance criteria to $Q = \text{diag}([280, 280, 210])$, $R = 0.05$, and $\gamma = 200$, with choosing C as an identity matrix to match with the provided full-state reference by the target system. Figure 6.5 illustrates the evolution of the controlled system and the target trajectories together with the control while learning process. In Figure 6.6, the tracking value and its parameters are shown. A video of the simulation showing the synchronization details is uploaded on https://youtu.be/1SnvDyb_7Os.

6.5 Summary

This chapter introduces an online learning-based method for nonlinear tracking with unknown dynamics. We assumed nonlinear control-affine dynamics

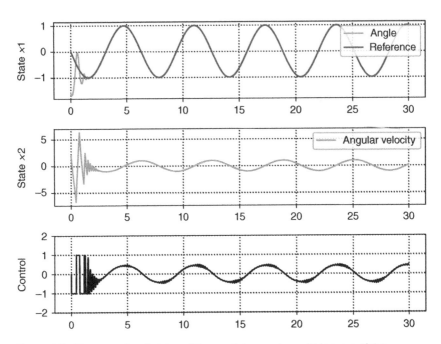

Figure 6.3 The control and states of the pendulum system within a run of the implemented learning approach from a randomly chosen initial condition where only the angle trajectory is provided as reference to be tracked.

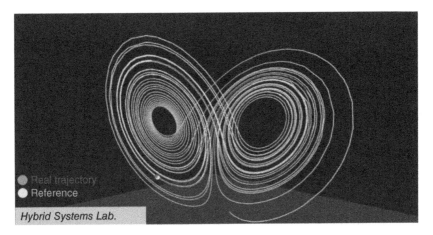

Figure 6.4 A view of the 3D simulation done for synchronizing the chaotic Lorenz system. The video can be accessed at: https://youtu.be/1SnvDyb_7Os.

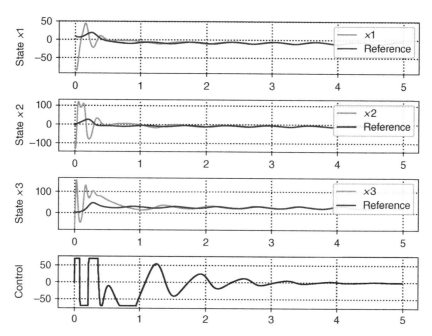

Figure 6.5 The states and the obtained control of the Lorenz system while learning to synchronize with the given reference trajectories, starting from a random initial condition.

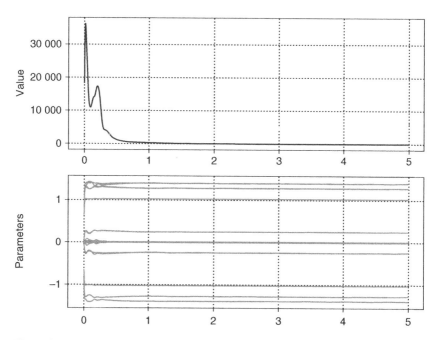

Figure 6.6 The evolutions of the value and parameters while learning the tracking controller of the Lorenz system, corresponding to Figure 6.5.

structured in terms of a set of bases functions. Accordingly, we formulated an optimal tracking control, where the objective function was redefined to conform with the structured system. This performance measure and a value function parameterized in a quadratic form were approximately minimized by solving a derived matrix differential equation. Hence, the formulation allows us to compose a learning-based tracking control framework, which relies only on the online measurements of the system states and the reference trajectory. In the simulation results, the proposed learning approach demonstrated satisfactory tracking of fully or partially provided reference trajectories. Considering the improved computational complexity of the obtained update rule for the value parameters, future research will include addressing practical limitations, and accordingly obtain an end-to-end platform for real-world applications.

Bibliography

Shubhendu Bhasin, Rushikesh Kamalapurkar, Marcus Johnson, Kyriakos G. Vamvoudakis, Frank L. Lewis, and Warren E. Dixon. A novel actor–critic–identifier architecture for approximate optimal control of uncertain nonlinear systems. *Automatica*, 49(1):82–92, 2013.

Steven L. Brunton, Joshua L. Proctor, and J. Nathan Kutz. Discovering governing equations from data by sparse identification of nonlinear dynamical systems. *Proceedings of the National Academy of Sciences of the United States of America*, 113(15):3932–3937, 2016.

Milad Farsi and Jun Liu. Structured online learning-based control of continuous-time nonlinear systems. *IFAC-PapersOnLine*, 53(2):8142–8149, 2020.

Milad Farsi and Jun Liu. A structured online learning approach to nonlinear tracking with unknown dynamics. In *Proceedings of the American Control Conference*, pages 2205–2211. IEEE, 2021.

Eurika Kaiser, J. Nathan Kutz, and Steven L. Brunton. Sparse identification of nonlinear dynamics for model predictive control in the low-data limit. *Proceedings of the Royal Society A*, 474(2219):20180335, 2018.

Rushikesh Kamalapurkar, Joel A. Rosenfeld, and Warren E. Dixon. Efficient model-based reinforcement learning for approximate online optimal control. *Automatica*, 74:247–258, 2016.

Anna Prach, Ozan Tekinalp, and Dennis S. Bernstein. Infinite-horizon linear-quadratic control by forward propagation of the differential Riccati equation. *IEEE Control Systems Magazine*, 35(2):78–93, 2015.

David Scherer, Paul Dubois, and Bruce Sherwood. VPython: 3D interactive scientific graphics for students. *Computing in Science & Engineering*, 2(5):56–62, 2000.

Huaguang Zhang, Lili Cui, Xin Zhang, and Yanhong Luo. Data-driven robust approximate optimal tracking control for unknown general nonlinear systems using adaptive dynamic programming method. *IEEE Transactions on Neural Networks*, 22(12):2226–2236, 2011.

Yuanheng Zhu, Dongbin Zhao, and Xiangjun Li. Using reinforcement learning techniques to solve continuous-time non-linear optimal tracking problem without system dynamics. *IET Control Theory and Applications*, 10(12):1339–1347, 2016.

7

Piecewise Learning and Control with Stability Guarantees

In this chapter, we extend the Structured Online Learning (SOL) framework to allow use of a more flexible piecewise parameterized model. The goal is to improve computational complexity, while retaining flexibility in learning. We would also like to provide closed-loop stability analysis of the unknown system and offer stability guarantees with learning. To this end, the piecewise learning framework lends itself to efficient verification using optimization-based techniques. The results presented in this chapter appeared in Farsi et al. [2022].

7.1 Introduction

In Chapter 5, we employed a set of bases to parameterize the system model. For this purpose, a set of polynomial or trigonometric bases is proven to be effective in approximating different functions with any arbitrary accuracy over a compact domain. Even though a set of polynomial bases, for instance, is known to be sufficient as a universal approximator, the number of bases required for a tight approximation of the dynamics over a given domain may be exceedingly high. The number of the bases, in fact, depends on the domain of interest, where a larger domain may exhibit nonlinearites that require a larger set of bases. This highly impedes implementations, especially in an online learning and control setting.

In an alternative approach, instead of adding many bases to cover a large domain of interest, we divide the domain into pieces where each can be handled independently with a limited number of bases. Employing a piecewise model will improve learning greatly by keeping the online computations needed for updating the model in a tractable size.

Despite the improvement in the computations, data efficiency of learning may be diminished if a large number of the pieces are chosen. Considering that the total number of the model parameters is relative to the number of pieces, a piecewise model may involve more parameters compared to learning in terms of bases.

Model-Based Reinforcement Learning: From Data to Continuous Actions with a Python-based Toolbox,
First Edition. Milad Farsi and Jun Liu.

In fact, there exists a trade-off between the data efficiency and computational efficiency that can be controlled by the number of pieces employed.

The rest of the chapter is presented in the following order. Section 7.2 formulates the problem. In Section 7.3, we propose a piecewise learning and control framework, where we first obtain an estimation of the system and then solve an approximate optimal control in a closed-loop form. In Section 7.4, we provide an upper bound for the uncertainty in the identified piecewise model based on the observations. In Section 7.5, the obtained uncertainty bounds are implemented to synthesize a Lyapunov function for the closed-loop system. In Section 7.6, two benchmark examples are discussed to numerically validate the approach.

7.2 Problem Formulation

Consider the nonlinear system in control-affine form

$$\dot{x} = F(x, u) = f(x) + g(x)u = f(x) + \sum_{j=1}^{m} g_j(x)u_j, \tag{7.1}$$

where $x \in D \subseteq \mathbb{R}^n$, $u \in U \subseteq \mathbb{R}^m$, $f : D \to \mathbb{R}^n$, and $g : D \to \mathbb{R}^{n \times m}$.

The cost functional to be minimized along the trajectory, started from the initial condition $x(0) = x_0$, is considered to be in the following linear quadratic form:

$$J(x_0, u) = \lim_{T \to \infty} \int_0^T e^{-\gamma t} \left(x^T Q x + u^T R u \right) dt, \tag{7.2}$$

where $Q \in \mathbb{R}^{n \times n}$ is positive semidefinite, $\gamma \geq 0$ is the discount factor, and $R \in \mathbb{R}^{m \times m}$ is a diagonal matrix with only positive values, given by the design criteria.

7.3 The Piecewise Learning and Control Framework

We approximate the nonlinear system (7.1) by a piecewise model with a bounded uncertainty:

$$\dot{x} = W_\sigma \Phi(x) + \sum_{j=1}^{m} W_{j\sigma} \Phi(x) u_j + d_\sigma, \tag{7.3}$$

where $d_\sigma \in \mathbb{R}^n$ is a time-varying uncertainty, W_σ and $W_{j\sigma} \in \mathbb{R}^{n \times p}$ are the matrices of the coefficients for $\sigma \in \{1, 2, \ldots, n_\sigma\}$ and $j \in \{1, 2, \ldots, m\}$, with a set of differentiable bases $\Phi(x) = [\phi_1(x) \quad \ldots \quad \phi_p(x)]^T$, and n_σ denoting the total number of pieces. Moreover, any piece of the system is defined over a convex set given by a set of linear inequalities as follows: $\Upsilon_\sigma = \{x \in D | Z_\sigma x \leq z_\sigma\}$, where $\sigma \in \{1, \ldots, n_\sigma\}$ and Z_σ and z_σ are a matrix and a vector of appropriate dimensions.

We assume that the set $\{\Upsilon_\sigma\}$ forms a partition of the domain and its elements do not share any interior points, i.e. $\bigcup_{\sigma=1}^{n_\sigma} \Upsilon_\sigma = D$ and $\text{int}[\Upsilon_\sigma] \cap \text{int}[\Upsilon_l] = \emptyset$ for $\sigma \neq l$ and $\sigma, l \in \{1, 2, \ldots, n_\sigma\}$. Furthermore, the piecewise model is assumed to be continuous across the boundaries of $\{\Upsilon_\sigma\}$ that will be discussed later in detail. The control input and the uncertainty are assumed to be bounded and lie in the sets: $U = \{u \in \mathbb{R}^m \||u_j| \leq \bar{u}_j, \forall j \in \{1, 2, \ldots, m\}\}$ and $\Delta_\sigma = \{d_\sigma \in \mathbb{R}^n \||d_{\sigma i}| \leq \bar{d}_{\sigma i}, \forall i \in \{1, 2, \ldots, n\}\}$, respectively. The uncertainty upper bound $\bar{d}_\sigma = (\bar{d}_{\sigma 1}, \ldots, \bar{d}_{\sigma n})$ is to be determined.

7.3.1 System Identification

Having defined the parameterized model of the system, we employ a system identification approach to update the system parameters. For each pair of samples obtained from the input and state of the system, i.e. (x^s, u^s), we first locate the element in the partition $\{\Upsilon_\sigma\}$ that contains the sampled state x^s. Then, we locally update the system coefficients of the particular piece from which the state is sampled. The weights are updated according to

$$[\hat{W}_\sigma \ \hat{W}_{1\sigma} \ \ldots \ \hat{W}_{m\sigma}]_k = \arg\min_{\bar{W}} \|\dot{X}_{k\sigma} - \bar{W}\Theta_{k\sigma}\|_2^2, \tag{7.4}$$

where k is the time step, and $\Theta_{k\sigma}$ includes a matrix of samples with

$$\Theta_k^s = [\Phi^T(x^s) \ \Phi^T(x^s)u_1^s \ \ldots \ \Phi^T(x^s)u_m^s]_k^T,$$

for the sth sample in the σth partition. Correspondingly, $\dot{X}_{k\sigma}$ contains the sampled state derivatives. While in principle, any identification technique can be used, e.g. Brunton et al. [2016] and Yuan et al. [2019], the linearity with respect to the coefficients allows us to employ least-square techniques. In this chapter, since an online application is intended, we implement the recursive least-square technique that provides a more computationally efficient way to update the parameters.

Continuity of the Identified Model
Considering that differentiable bases are assumed, the model identified is differentiable within the interior of Υ_σ for $\sigma \in \{1, 2, \ldots, n_\sigma\}$. However, the pieces of the model may not meet in the boundaries of Υ_σ, where $x \in \Upsilon_\sigma \cap \Upsilon_l$ for any $\sigma \neq l$ and $\sigma, l \in \{1, 2, \ldots, n_\sigma\}$.

Based on our knowledge of system (7.1) from which we collect samples, the continuity holds for the original system. Hence, in theory, if many pieces are chosen, and enough samples are collected, the edges of pieces will converge together to yield a continuous model. However, choosing arbitrarily, small pieces are not practical.

There exist different techniques to efficiently choose the partitions on D and best fit a continuous piecewise model, see, e.g. Toriello and Vielma [2012], Breschi

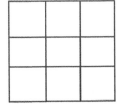 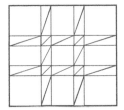

Figure 7.1 A scheme of obtaining a continuous piecewise model is illustrated. On the left, partitions on two-dimensional domains are shown for which the pieces of the model may not be connected in the borders. On the right, some extra triangular pieces are constructed to allow filling the possible gaps in the model.

et al. [2016], Ferrari-Trecate et al. [2003], Amaldi et al. [2016], and Rebennack and Krasko [2020]. Such techniques usually involve global adjustments of the model weights and the partitions for which the computations can be considerably expensive. Therefore, we choose to locally deal with the gaps among the pieces. This can be done by a postprocessing routine performed on the identified model.

A rather straightforward technique is to define extra partitions in the margins of each Υ_σ to fill the gaps among pieces. The weights of the corresponding pieces added can be chosen according to the weights of the adjacent pieces that is given by the identification. This is done in a way that they help to connect all the pieces together to make a continuous piecewise model. Figure 7.1 illustrates the process of constructing extra partitions for a two-dimensional case, where we choose them to be in triangular shapes. A similar approach can be taken for generalizing to the n-dimensional case.

7.3.2 Database

Although an online technique is used to update the piecewise model along trajectories, we still need to collect a number of samples for each piece of the system. The set of samples recorded will be used later to obtain an estimation of the uncertainty bounds for each mode of the system. For this purpose, we, over time, hand pick and save samples that best describe the dynamics in any mode of the piecewise system.

It should be noted that the database will be processed offline to extract the uncertainty bounds. Hence, it does not affect the online learning procedure and its computational cost. Any sample of the system, to be stored in the database, includes $(\Theta_k^s, \hat{\dot{x}}_k)$, where the state derivative is approximated by $\hat{\dot{x}}_k = (x_k - x_{k-1})/h$ and $\dot{e}_k$. For better results, higher-order approximations of the state derivative can be employed.

Different techniques can be employed to obtain a summary of the samples collected. We assume a given maximum size of database N_{d}. Then, for any mode of the piecewise model, we keep adding the samples with larger prediction errors to the database. Therefore, at any step, we compare the prediction error $\dot{e}_k = \|\dot{x}_k - \hat{\dot{x}}_k\|$ with the most recent average error $\bar{e}_{k\sigma}$ obtained for the active piece. Hence, if the condition $\dot{e}_k > \eta \bar{e}_{k\sigma}$ holds, we add the sample to the database, where the constant $\eta > 0$ adjusts the threshold. If the maximum number of samples in database is reached, we replace the oldest sample with the recent one.

7.3.3 Feedback Control

In Chapter 5, a matrix differential equation is proposed using a quadratic parametrization in terms of the bases functions to obtain a feedback control. Here, we adopt a similar learning framework, but consider a family of n_σ differential equations, each of which corresponds to one particular mode of the system in the piecewise model. We integrate the following state-dependent Riccati differential equation in forward time:

$$-\dot{P}_\sigma = \bar{Q} + P_\sigma \frac{\partial \Phi(x)}{\partial x} W_\sigma + W_\sigma^T \frac{\partial \Phi(x)}{\partial x}^T P_\sigma - \gamma P_\sigma$$
$$- P_\sigma \frac{\partial \Phi(x)}{\partial x} \left(\sum_{j=1}^{m} W_{j\sigma} \Phi(x) r_j^{-1} \Phi(x)^T W_{j\sigma}^T \right) \frac{\partial \Phi(x)}{\partial x}^T P_\sigma. \tag{7.5}$$

The solution to the differential equation (7.5) characterizes the value function defined by

$$V_\sigma = \Phi^T P_\sigma \Phi, \tag{7.6}$$

based on which we obtain a piecewise control

$$u_j = -r_j^{-1} \frac{\partial V_\sigma}{\partial x}^T g_j(x) = -\Phi(x)^T r_j^{-1} P_\sigma \frac{\partial \Phi(x)}{\partial x} W_{j\sigma} \Phi(x). \tag{7.7}$$

7.4 Analysis of Uncertainty Bounds

We use the uncertainty in the piecewise system (7.3) to capture approximation errors in identification. In this section, we analyze the worst-case bounds to provide guarantees for the proposed framework.

There exist two sources of uncertainty that affect the accuracy of the identified model. The first is the mismatch between the identified model and the observations made. The latter may also be affected by the measurement noise. The second is due to unsampled areas in the domain. We can estimate the uncertainty bound for any piece of the model by combining these two bounds. In what follows, we discuss the procedure of obtaining these bounds in more detail.

Assumption 7.1 *For any given* (x^s, u^s), *let* $F_i(x^s, u^s)$ *be the ith element of* $F(x^s, u^s)$. *We assume that* $F_i(x^s, u^s)$ *can be measured with some tolerance as* $\tilde{F}_i(x^s, u^s)$, *where* $|\tilde{F}_i(x^s, u^s) - F_i(x^s, u^s)| \leq \varrho_e |\tilde{F}_i(x^s, u^s)|$ *with* $0 \leq \varrho_e < 1$ *for all* $i \in \{1, \ldots, n\}$.

We make predictions $\hat{F}_i(x^s, u^s)$ of the state derivatives for any sample using the identified model. Hence, we can easily compute the distance between the prediction and the approximate evaluation of the system by using the samples collected for any piece. This gives the loss $|\hat{F}_i(x^s, u^s) - \tilde{F}_i(x^s, u^s)|$.

Theorem 7.1 *Let Assumption 7.1 hold, and* S_{Υ_σ} *denotes the set of indices for sample pairs* (x^s, u^s) *such that* $x^s \in \Upsilon_\sigma$. *Then, an upper bound of the prediction error, regarding any sample* (x^s, u^s) *for* $s \in \{1, \ldots, N_s\}$, *is given by*

$$|\hat{F}_i(x^s, u^s) - F_i(x^s, u^s)| \leq \bar{d}_{e\sigma i} := \max_{s \in S_{\Upsilon_\sigma}} (|\hat{F}_i(x^s, u^s) - \tilde{F}_i(x^s, u^s)| + \varrho_e |\tilde{F}_i(x^s, u^s)|),$$

where $\sigma \in \{1, \ldots, n_\sigma\}$, *and* $i \in \{1, \ldots, n\}$.

Proof: According to Assumption 7.1, it is straightforward to show that the prediction error can be bound for any σ by using the samples in partition σ as follows:

$$|\hat{F}_i(x^s, u^s) - F_i(x^s, u^s)| \leq |\hat{F}_i(x^s, u^s) - \tilde{F}_i(x^s, u^s)| + |\tilde{F}_i(x^s, u^s) - F_i(x^s, u^s)|$$

$$\leq |\hat{F}_i(x^s, u^s) - \tilde{F}_i(x^s, u^s)| + \varrho_e |\tilde{F}_i(x^s, u^s)|$$

$$\leq \max_{s \in S_{\Upsilon_\sigma}} (|\hat{F}_i(x^s, u^s) - \tilde{F}_i(x^s, u^s)| + \varrho_e |\tilde{F}_i(x^s, u^s)|)$$

$$= \bar{d}_{e\sigma i}. \qquad \blacksquare$$

7.4.1 Quadratic Programs for Bounding Errors

The samples may not be uniformly obtained from the domain. Depending on how smooth the dynamics are, there might be unpredictable behavior of the system in the gaps among the samples. Hence, the predictions made by the identified model may be misleading in the areas we have not visited yet. To take this into account, we assume that a Lipschitz constant is given for the system. More specifically, we let $\varrho_x \in \mathbb{R}^n_+$ and $\varrho_u \in \mathbb{R}^n_+$ denote the Lipschitz constants of $F(x, u)$ with respect to x and u on $D \times U$, respectively. We use this to bound the uncertainty for the unsampled areas.

We need to compute the worst-case of the prediction error within any piece that is given by $|\hat{F}_i(x, u) - F_i(x, u)|$, where $\hat{F}(\cdot, \cdot)$ denotes an evaluation of the identified model. However, according to Assumption 7.1, we do not have access to the original system to exactly evaluate $F(\cdot, \cdot)$. Therefore, we obtain the bound in terms of the approximate value instead.

Assumption 7.2 *For system (7.1), $\exists \varrho_x \in \mathbb{R}_+^n$ such that we have*

$$|F_i(x_0, u) - F_i(y_0, u)| \leq \varrho_{xi} \|x_0 - y_0\|,$$

for any $x_0, y_0 \in D$, and $u \in U$, where $i \in \{1, \dots, n\}$.

Assumption 7.3 *For system (7.1), $\exists \varrho_u \in \mathbb{R}_+^n$ such that we have*

$$|F_i(x, u_0) - F_i(x, w_0)| \leq \varrho_{ui} \|u_0 - w_0\|,$$

for any $x \in D$, and $u_0, w_0 \in U$, where $i \in \{1, \dots, n\}$.

Assumption 7.4 *An initial estimation of ϱ_e and Lipschitz constants ϱ_{xi} and ϱ_{ui} is known.*

The following results and the bounds will directly depend on the choice of ϱ_x, and ϱ_u. However, this is the least we can assume that allows us to carry out the computations. Moreover, making such assumptions is not restrictive in practice since we often have a general knowledge of the application. Moreover, the learning may be first started with an initial guess of the continuity constants. Later, if the samples collected override the assumption made, we can update these values.

To calculate the uncertainty bound for any piece, we first look for the largest gap existing among the samples within each piece. The procedure starts with searching for the largest gaps in the state and control spaces that do not contain any samples as show in Figure 7.2. Let (x^{s*}, u^{s*}) be the closest sample indexed in S_{Υ_σ} to the center point $(c_{x\sigma}^*, c_{u\sigma}^*)$ of the sample gap (as a Euclidean ball) with radius $(r_{x\sigma}^*, r_{u\sigma}^*)$ (Figure 7.3). For this purpose, we solve a quadratic programming (QP) problem for each piece. The solution to the following QP returns the center $c_{x\sigma}^*$ at which an n-dimensional ball of the largest radius $r_{x\sigma}^*$ can be found in the σth piece such that no samples x^s are contained in this ball:

$$\underset{c_{x\sigma}, r_{x\sigma}}{\arg\max} \quad r_{x\sigma} \tag{7.8}$$

$$\text{subject to} \quad c_{x\sigma} \in \Upsilon_\sigma$$

$$\text{for} \quad s \in S_{\Upsilon_\sigma} : \quad \|x^s - c_{x\sigma}\| \geq r_{x\sigma}.$$

Similarly, we can obtain the center $c_{u\sigma}^*$ and radius $r_{u\sigma}^*$ to represent the sample gap as an m-dimensional ball in the control space by solving

$$\underset{c_{u\sigma}, r_{u\sigma}}{\arg\max} \quad r_{u\sigma} \tag{7.9}$$

$$\text{subject to} \quad c_{u\sigma} \in U$$

$$\text{for} \quad s \in S_{\Upsilon_\sigma} : \quad \|u^s - c_{u\sigma}\| \geq r_{u\sigma}.$$

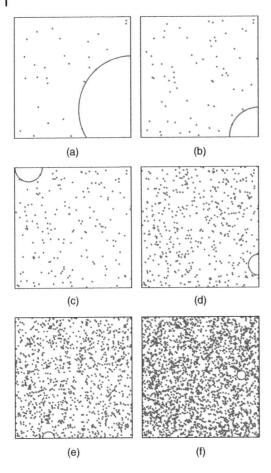

Figure 7.2 Subfigures (a)–(f) denote the sample gaps located for different number of samples. It is observed that the radius of the gap decreases by increasing the number of the samples.

(a)   (b)

(c)   (d)

(e)   (f)

Theorem 7.2 *Let Assumptions 7.1–7.4 hold and* $(r_{x\sigma}^*, r_{u\sigma}^*)$ *is given by the solutions of (7.8) and (7.9). Then, an upper bound for the prediction error can be obtained regarding all unvisited points* $x \in \Upsilon_\sigma$ *and* $u \in U$ *as follows:*

$$|F_i(x,u) - \hat{F}_i(x,u)| \leq \bar{d}_{\sigma i} = \varrho_{ui} r_{u\sigma}^* + \varrho_{xi} r_{x\sigma}^* + \bar{d}_{e\sigma i} + \hat{\varrho}_{ui} r_{u\sigma}^* + \hat{\varrho}_{xi} r_{x\sigma}^*. \quad (7.10)$$

Proof: According to the Lipschitz condition, the following holds for any $(x,u) \in \Upsilon_\sigma$

$$|F_i(x,u) - F_i(x^{s*}, u^{s*})|$$

$$\leq |F_i(x,u) - F_i(x, u^{s*})| + |F_i(x, u^{s*}) - F_i(x^{s*}, u^{s*})|$$

$$\leq \varrho_{ui} \|u - u^{s*}\| + \varrho_{xi} \|x - x^{s*}\| . \quad (7.11)$$

Figure 7.3 The scheme for obtaining the uncertainty bound according to the sample gap. Black dots denote the measurements.

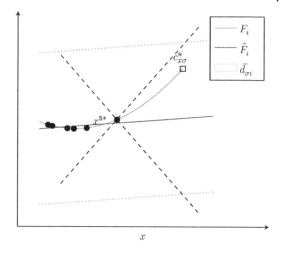

Moreover, we have the estimation $\hat{F}(x, u)$ of the system. Then, the difference is bounded by

$$|F_i(x, u) - \hat{F}_i(x, u)|$$

$$\leq |F_i(x, u) - F_i(x^{s*}, u^{s*})| + |F_i(x^{s*}, u^{s*}) - \hat{F}_i(x, u)|$$

$$\leq |F_i(x, u) - F_i(x^{s*}, u^{s*})| + |F_i(x^{s*}, u^{s*}) - \hat{F}_i(x^{s*}, u^{s*})|$$
$$\quad + |\hat{F}_i(x^{s*}, u^{s*}) - \hat{F}_i(x, u)|$$

$$\leq \varrho_{ui} \|u - u^{s*}\| + \varrho_{xi} \|x - x^{s*}\| + \bar{d}_{e\sigma i} + |\hat{F}_i(x^{s*}, u^{s*}) - \hat{F}_i(x, u)|$$

$$\leq \varrho_{ui} \|u - u^{s*}\| + \varrho_{xi} \|x - x^{s*}\| + \bar{d}_{e\sigma i} + |\hat{F}_i(x^{s*}, u^{s*}) - \hat{F}_i(x^{s*}, u)|$$
$$\quad + |\hat{F}_i(x^s, u) - \hat{F}_i(x, u)|$$

$$\leq \varrho_{ui} \|u - u^{s*}\| + \varrho_{xi} \|x - x^{s*}\| + \bar{d}_{e\sigma i} + \hat{\varrho}_{ui} \|u - u^{s*}\| + \hat{\varrho}_{xi} \|x - x^{s*}\|,$$

where in the last step, we used inequality (7.11) and the bound obtained in Theorem 7.1 according to the samples. Then, considering that $\hat{F}_i(x, u)$ is known, we can easily compute the corresponding Lipschitz constants $\hat{\varrho}_{ui}$ and $\hat{\varrho}_{xi}$. The largest distance with the closest sample (x^{s*}, u^{s*}) happens in the sample gap given with the radius $r^*_{x\sigma}$, and $r^*_{u\sigma}$. This yields the total bound of the error as follows:

$$|F_i(x, u) - \hat{F}_i(x, u)| \leq \varrho_{ui} r^*_{u\sigma} + \varrho_{xi} r^*_{x\sigma} + \bar{d}_{e\sigma i} + \hat{\varrho}_{ui} r^*_{u\sigma} + \hat{\varrho}_{xi} r^*_{x\sigma} \qquad \blacksquare$$

7.5 Stability Verification for Piecewise-Affine Learning and Control

7.5.1 Piecewise Affine Models

A special case of system (7.3) can be obtained when we choose $\Phi(x) = \begin{bmatrix} 1 & x^T \end{bmatrix}$. We consider system coefficients in the form of $W_\sigma = \begin{bmatrix} C_\sigma & A_\sigma \end{bmatrix}$ and $W_{j\sigma} = \begin{bmatrix} B_{j\sigma} & 0 \end{bmatrix}$.

Clearly, A_σ, $B_{j\sigma}$, and C_σ can be used to rewrite the Piecewise Affine (PWA) system in the standard form

$$\dot{x} = A_\sigma x + \sum_{j=1}^{m} B_{j\sigma} u_j + C_\sigma + d_\sigma. \tag{7.12}$$

7.5.2 MIQP-based Stability Verification of PWA Systems

In this section, we adopt an mixed-integer quadratic program (MIQP)-based verification technique based on the approach presented in Chen et al. [2020]. In this framework, by considering a few steps ahead, we verify that the Lyapunov function is decreasing. However, it may not be necessarily monotonic, meaning that it may be increasing in some steps and then be decreasing greatly in some other steps to compensate. Regarding the fact that this approach is inherently a discrete technique, we need to consider a discretization of (7.12). By Euler approximation, we have

$$x_{k+1} = \check{F}_{\mathrm{d}}(x_k, u_k) = \check{A}_\sigma x_k + \sum_{j=1}^{m} \check{B}_{j\sigma} u_{jk} + \check{C}_\sigma + d_\sigma, \tag{7.13}$$

where $\check{A}_\sigma$, $\check{B}_{j\sigma}$, and $\check{C}_\sigma$ are the discrete system matrices of the same dimension as (7.12). Moreover, we re-adjust the uncertainty bound as $\bar{d}_\sigma := h\bar{d}_\sigma$, where h denotes the time step.

We refer to the uncertain closed loop system with the control $u_{jk} = \omega_j(x_k)$ as

$$x_{k+1} = \check{F}_{\mathrm{d,cl}}(x_k). \tag{7.14}$$

For this system, let the convex set $\bar{D} = \{x \in D | Z_{\bar{D}} x \leq z_{\bar{D}}\}$ be a user-defined region of interest, within which obtaining a region of attraction (ROA) is desirable.

Searching for a Lyapunov Function

We summarize an altered version of the technique for obtaining a Lyapunov function that is first presented in Chen et al. [2020] for a deterministic closed-loop system with the neural network controller. Hence, we modify the algorithm to allow the uncertainty together with the feedback control (7.7).

The procedure includes two stages that are performed iteratively until a Lyapunov function is obtained and verified, or it is concluded that there exist no Lyapunov functions in the given set of candidates.

In the first stage, we assume an initial set of Lyapunov candidates in the form of (7.15). Then, the learner searches for a subset for which the negativity of the Lyapunov difference can be guaranteed with respect to a set of samples collected from the system. If such subset exists, one element in this subset is proposed as the Lyapunov candidate by the learner.

In the second stage, the proposed Lyapunov candidate is verified on the original system. Noting that the learner only uses a finite number of samples for suggesting a Lyapunov candidate, it may not be valid for all the evolutions of the uncertain system. Accordingly, the verifier either certifies the Lyapunov candidate, or finds a point as the counterexample for which the Lyapunov candidate fails. This sample is added to the set of samples collected from the system. Then, we again proceed to the learner stage with the updated set of samples.

The algorithm is run in a loop, where we start with an empty set of samples in the learner. Then, we continue with proposing a Lyapunov candidate, and adding one counterexample in each iteration of the loop. While growing the set of samples, the set of Lyapunov candidates shrinks in every iteration until it is either validated, or no element is left in the set meaning that no such Lyapunov exists.

Learning and Verification of a Lyapunov Function

Assuming $u_j = -r_j^{-1} B_{j\sigma}^T P_{3\sigma} x_k$, and defining $\check{A}_{\mathrm{cl},\sigma} = \check{A}_\sigma - \sum_{j=1}^m r_j^{-1} \check{B}_{j\sigma} B_{j\sigma}^T P_\sigma$, the discrete closed-loop system becomes $x_{k+1} = \check{A}_{\mathrm{cl},\sigma} x_k + \check{C}_\sigma + d_\sigma$.

Now, consider the Lyapunov function:

$$V(x_k, \hat{P}) = \begin{bmatrix} x_k \\ x_{k+1} \end{bmatrix}^T \hat{P} \begin{bmatrix} x_k \\ x_{k+1} \end{bmatrix} \tag{7.15}$$

characterized by $\hat{P} \in \mathscr{F}$, where

$$\mathscr{F} = \{\hat{P} \in \mathbb{R}^{2n \times 2n} | 0 \le \hat{P} \le I, V(x_{k+1}, \hat{P}) - V(x_k, \hat{P}) < 0,$$

$$\forall x_k \in \bar{D} \backslash \{0\}, d_\sigma \in \Delta_\sigma\}.$$

The structure of the Lyapunov function is suggested by Chen et al. [2020] that employs a piecewise quadratic function to parameterize the Lyapunov function. This approach combines the nonmonotonic Lyapunov function [Ahmadi and Parrilo, 2008] and finite-step Lyapunov function [Bobiti and Lazar, 2016, Aeyels and Peuteman, 1998] techniques to provide a guarantee by looking at the next few steps. It should be noted that the Lyapunov function may not be necessarily decreasing within any single step, while it must be decreasing within the finite steps taken into account.

The Learner   To realize a Lyapunov function, one needs a mechanism to look for the appropriate values of $\hat{P}$ within $\mathscr{F}$. For this purpose, we obtain an overapproximation of $\mathscr{F}$ by considering only finite number of elements in $(\bar{D}, \Delta)$. Let us first define the increment on the Lyapunov function as

$$\Delta V(x, \hat{P}) = V(\check{F}_{\mathrm{d,cl}}(x), \hat{P}) - V(x, \hat{P})$$

$$= \begin{bmatrix} \check{F}_{\mathrm{d,cl}}(x) \\ \check{F}_{\mathrm{d,cl}}^{(2)}(x) \end{bmatrix}^T \hat{P} \begin{bmatrix} \check{F}_{\mathrm{d,cl}}(x) \\ \check{F}_{\mathrm{d,cl}}^{(2)}(x) \end{bmatrix} - \begin{bmatrix} x \\ \check{F}_{\mathrm{d,cl}}(x) \end{bmatrix}^T \hat{P} \begin{bmatrix} x \\ \check{F}_{\mathrm{cl}}(x) \end{bmatrix},$$

where $\check{F}_{\mathrm{d,cl}}^{(2)}(x) = \check{F}_{\mathrm{d,cl}}(\check{F}_{\mathrm{d,cl}}(x))$.

Furthermore, assume that the set of N_s number of samples are given as follows:

$$\mathcal{S} = \{(x, \breve{F}_{d,cl}(x), \breve{F}^{(2)}_{d,cl}(x))_1, \dots, (x, \breve{F}_{d,cl}(x), \breve{F}^{(2)}_{d,cl}(x))_{N_s}\}.$$

Note that $\mathcal{S}$ implicitly includes samples of the disturbance input and the state. Now, using $\mathcal{S}$, we obtain the overapproximation

$$\tilde{\mathcal{F}} = \{\hat{P} \in \mathbb{R}^{2n \times 2n} | 0 \leq \hat{P} \leq I, \Delta V(x, \hat{P}) \leq 0, \forall x \in \mathcal{S}, d_\sigma \in \Delta_\sigma\}.$$

To find an element in $\tilde{\mathcal{F}}$, there exist efficient iterative techniques that are well known as cutting-plane approaches. See, e.g. Atkinson and Vaidya [1995], Elzinga and Moore [1975], and Boyd and Vandenberghe [2007]. In Chen et al. [2020], the Analytic Center Cutting-Plane Method (ACCPM) [Goffin and Vial, 1993, Nesterov, 1995, Boyd and Vandenberghe, 2004] is employed in an optimization problem:

$$\hat{P}^{(i)} = \arg\min_{\hat{P}} - \sum_{x \in \mathcal{S}_i} \log(-\Delta V(x, \hat{P})) - \log \det (I - \hat{P}) - \log \det(\hat{P}), \quad (7.16)$$

where i is the iteration index. If feasible, the log-barrier function in the first term guarantees the solution within $\tilde{\mathcal{F}}$ for which the negativity of the Lyapunov difference holds. The other two terms ensure $0 \leq \hat{P}^{(i)} \leq I$. The solution gives a Lyapunov function V based on the set of the samples $\mathcal{S}_i$ in the ith stage. On the other hand, if a solution does not exist, the set $\mathcal{F}$ is concluded to be empty.

The Verifier The Lyapunov function candidate suggested by (7.16) may not guarantee asymptotic stability for all $x \in \bar{D}$ and $d_\sigma \in \Delta_\sigma$ since only the sampled space was considered. Therefore, in the next step, we need to verify the Lyapunov function candidate for the uncertain system. To do so, a MIQP is solved based on the convex hull formulation of the PWA:

$$\max_{x^j, u^j, d^j, \mu^j} \begin{bmatrix} x^1 \\ x^2 \end{bmatrix}^T \hat{P}^{(i)} \begin{bmatrix} x^1 \\ x^2 \end{bmatrix} - \begin{bmatrix} x^0 \\ x^1 \end{bmatrix}^T \hat{P}^{(i)} \begin{bmatrix} x^0 \\ x^1 \end{bmatrix} \quad (7.17)$$

subject to

$$Z_{\bar{D}} x^0 \leq z_{\bar{D}}, \|x^0\|_\infty \geq \epsilon \quad (7.18)$$

$$u^j = \omega(x^j) \quad (7.19)$$

$$Z_\sigma x^j_\sigma \leq \mu^j_\sigma z_\sigma, Z_u u_\sigma \leq \mu^j_\sigma z_u, |d^j_{\sigma i}| \leq \mu^j_\sigma \bar{d}_{\sigma i}, \quad (7.20)$$

$$(1, x^j, u^j, d^j, x^{j+1})$$

$$= \sum_{\sigma=1}^{N_\sigma} (\mu^j_\sigma, x^j_\sigma, u^j_\sigma, d^j_\sigma, A_\sigma x^j_\sigma + B_\sigma u^j_\sigma + \mu^j_\sigma c_\sigma + d^j_\sigma) \quad (7.21)$$

$$\mu_\sigma \in \{0, 1\},$$

$$\forall \sigma \in \{1, \dots, N_\sigma\}, i \in \{1, \dots, n\}, j \in \{0, 1\}, \quad (7.22)$$

where a ball of radius ϵ around the origin is excluded from the set of states, and ϵ is chosen small enough in (7.18). This is due to Remark 7.1 and the fact that the numerical value of the objective becomes considerably small when approaching the origin. This makes the negativity of the objective too hard to verify around the origin. For more details in the implementation of the algorithm, we refer the reader to Chen et al. [2020].

The system is given by (7.21) and (7.22). To define the piecewise system in a mixed-integer problem, similar to Chen et al. [2020], we use the convex-hull formulation of piecewise model that is presented in Marcucci and Tedrake [2019]. However, to consider the uncertainty, we compose a slightly different system where we define extra variables to model the disturbance input.

Constraints (7.18) and (7.20) define the sets of the initial condition, the state, the control, and the disturbance inputs, respectively. Furthermore, the feedback control is implemented by (7.19).

To certify the closed-loop system as asymptotically stable, the optimal value returned by the Mixed-Integer Quadratic Program (MIQP) (7.17) is required to be negative. Otherwise, the argument (x^{0*}, x^{1*}, x^{2*}) of the optimal solution is added to the set of samples $\mathcal{S}$ as a counterexample.

7.5.3 Convergence of ACCPM

The convergence and complexity of the ACCPM for searching a quadratic Lyapunov function is discussed in Sun et al. [2002] and Chen et al. [2020], where an upper bound is obtained for the number of steps taken until the algorithm exits.

Lemma 7.1 *Let $\mathcal{F}$ be a convex subset of $\mathbb{R}^{n \times n}$. Moreover, there exists $P_{center} \in \mathbb{R}^{n \times n}$ such that $\{P \in \mathbb{R}^{n \times n} \mid \|P - P_{center}\|_F \leq \epsilon\} \subseteq \mathcal{F}$, where Frobenius norm is used, and $\mathcal{F} \subseteq \{P \in \mathbb{R}^{n \times n} \mid 0 \leq P \leq I\}$. Then, the center cutting-plane algorithm concludes in at most $O(n^3/\epsilon^2)$ steps.*

Proof: See Sun et al. [2002] and Chen et al. [2020] for the proof. ∎

Stability Analysis

Combining the uncertainty bounds in Section 7.4 and the Lyapunov-based verification results of this section, we are able to prove the following practical stability results of the closed-loop system.

Theorem 7.3 *Suppose that the MIQP (7.17) yields a negative optimal value. Let B_ϵ denote the set $\{x \in \mathbb{R}^n \mid \|x\|_\infty \leq \epsilon\}$, i.e. the ball of radius ϵ in infinity norm around the origin. Then the set B_ϵ is asymptotically stable for the closed-loop system (7.14). The largest sublevel set of V, i.e. $\{x \in \mathbb{R}^n \mid V(x) \leq c\}$ for some c, contained in $\bar{D}$ is a verified underapproximation of the real ROA.*

Proof: According to the conditions of the verifier, if the optimal value returned by the MIQP (7.17) is negative, we have effectively verified the following Lyapunov conditions:

$$V(0) = 0, \quad V(x) > 0, \quad \forall x \in \bar{D} \backslash \{0\}, \tag{7.23}$$

$$V(\check{F}_{\mathrm{d,cl}}(x)) - V(x) < 0, \quad \forall x \in \bar{D} \backslash B_\epsilon, d \in \Delta_\sigma, \tag{7.24}$$

for the uncertain closed-loop system (7.14). By standard Lyapunov analysis for set stability [Haddad and Chellaboina, 2011, Jiang and Wang, 2001], the set B_ϵ, which is the ball of radius ϵ in infinity norm around the origin, is asymptotically stable for system (7.14). Furthermore, any sublevel set of $V(x)$, i.e. $\{x \in \mathbb{R}^n \mid V(x) \leq c\}$ for some c, contained in $\bar{D}$ is contained in the ROA of B_ϵ. ∎

Remark 7.1 Due to the existence of a nonzero additive uncertainty bound, one cannot expect convergence to the origin precisely. This issue is addressed by providing convergence guarantee to a small neighborhood of the origin, i.e. B_ϵ. By collecting enough samples around the origin, a local approximation of the system is obtained by the mode $\sigma = 0$ of the identified system, whose domain includes the origin, while d_σ can be made arbitrarily small as $x_k \to 0$. By doing so, we can make ϵ in Theorem 7.3 arbitrarily small and the stability result is practically equivalent to the asymptotic stability of the origin. Alternatively, one can assume that there exists a local stabilizing controller that one can switch to when entering a small neighborhood of the origin. In this case, asymptotic stability can be achieved.

7.6 Numerical Results

To validate the proposed piecewise learning and verification technique, we implemented the approach on the pendulum system as (5.23) and the dynamical vehicle system [Pepy et al., 2006]. Moreover, we compared the results with other techniques presented in the literature. To make a fair comparison, we have taken the parameters of the system from Chang et al. [2019]. We performed all the simulations in Python 3.7 on a 2.6 GHz Intel Core i5 CPU.

7.6.1 Pendulum System

For the pendulum system, we discuss the simulation results in three sections. In the first section, we will explain the procedure of identifying the uncertain PWA model with a piecewise feedback control. In the second section, we verify the closed-loop uncertain system and obtain an ROA in $\bar{D}$. In the third section, we will present the comparison results.

Identify and Control

Control objective is to stabilize the pendulum at the top equilibrium point given by $x_{eq} = (0, 0)$. First, we start with learning a piecewise model together with the uncertainty bounds and the feedback control. For this purpose, we sample the system and update our model as discussed in Section 7.3.1. We set the sampling time as $h = 5$ ms. Accordingly, the value function and the control rule are updated online as in Section 7.3.3. Then, to verify the value to be decreasing within each mode, it only remains to calculate the uncertainty bounds using the results obtained in Section 7.4.

To make a visualization of the nonlinearity in the pendulum system (5.23) possible, we portray the second dynamic assuming $u = 0$ in Figure 7.4, where the first dynamic is only linear. The procedure of learning is illustrated through several stages in Figure 7.5. In the first column from the left, we illustrated the estimations only for the second dynamic with $u = 0$ to be comparable to Figure 7.4. Accordingly, it can be observed that the system identifier is able to closely approximate the nonlinearity with a piecewise model (Figure 7.6).

It should be noted, the learning is started from the mode containing the origin in its domain, that we label by $\sigma = 0$. As we collect more random samples in Υ_0, we can effectively decrease the uncertainty of the model around the origin and obtain a local controller. Then, we gradually expand the areas sampled to train the rest of the pieces in the PWA model.

Remark 7.2 It is worth mentioning that the model obtained and the uncertainty bounds can be further improved by continuing the sampling. In this

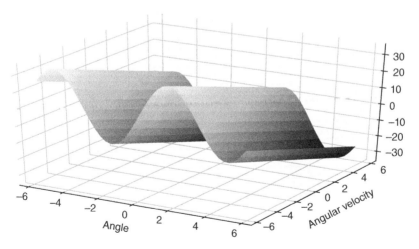

Figure 7.4 A view of the second dynamic of pendulum system (5.23) assuming $u = 0$ that is $f_2(x_1, x_2)$.

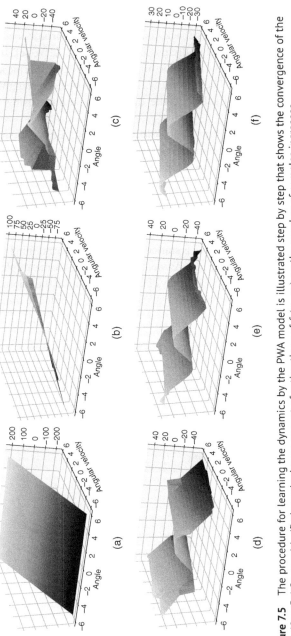

Figure 7.5 The procedure for learning the dynamics by the PWA model is illustrated step by step that shows the convergence of the identifier. Subfigures (a)–(f) show the improvement of estimations of $f_2(x_1, x_2)$, as the number of samples increases.

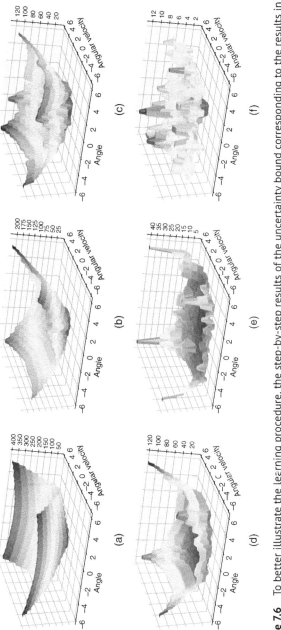

Figure 7.6 To better illustrate the learning procedure, the step-by-step results of the uncertainty bound corresponding to the results in Figure 7.5 are provided. It is evident that error bound is improved in every step.

implementation, we perform sampling only until the uncertainty bound obtained allows us to verify a decreasing value function for each piece of the PWA system.

Verification

Having the system identified and the feedback control, we can apply the verification algorithm based on MIQP problem. As done in Chen et al. [2020], we implemented the learner in CVXpy [Diamond and Boyd, 2016] with MOSEK [MOSEK ApS, 2020] solver, and the verifier in Gurobi 9.1.2 [Gurobi Optimization, 2020].

We choose $\bar{D}$ such that x_1 and $x_2 \in [-6, 6]$. To verify the system, we ran the algorithm and obtained a matrix $\hat{P}$ that characterizes the Lyapunov function as in (7.15).

$$\hat{P} = \begin{bmatrix} 0.69371067 & 0.02892586 & 0.1944487 & 0.05196313 \\ 0.02892586 & 0.26941371 & 0.02718769 & -0.21348358 \\ 0.1944487 & 0.02718769 & 0.69518109 & 0.05041737 \\ 0.05196313 & -0.21348358 & 0.05041737 & 0.33469316 \end{bmatrix}.$$

The largest level set of the associated Lyapunov function in $\bar{D}$ is pictured in Figure 7.8a as the ROA of the closed-loop system. Moreover, we illustrate different trajectories of the controlled system that confirms the verified Lyapunov function by constructing an ROA around the origin.

Comparison Results

To highlight the merits of the proposed piecewise learning approach, we compare the ROA obtained by different approaches in the literature. Chang et al. [2019] proposed a Neural Network (NN) Lyapunov function for stability verification. According to Chang et al. [2019], the comparison done on the pendulum system with Linear Quadratic Regulator (LQR) and Sum of Squares (SOS) showed noticeable superiority of the NN-based Lyapunov approach. Following the comparison results from Chang et al. [2019], we compare the ROA obtained by our approach with NN, SOS, and LQR techniques in Figure 7.8b. Clearly, the ROA obtained by the piecewise controller with the nonmonotonic Lyapunov function is considerably larger than the ones obtained by NN, SOL, and LQR algorithms as shown in Chang et al. [2019].

7.6.2 Dynamic Vehicle System with Skidding

In this section, to better demonstrate the merits of the algorithm proposed, we implemented the approach on a more complex system. The kinematic model of the vehicle system does not consider the real behavior of the system at high speed

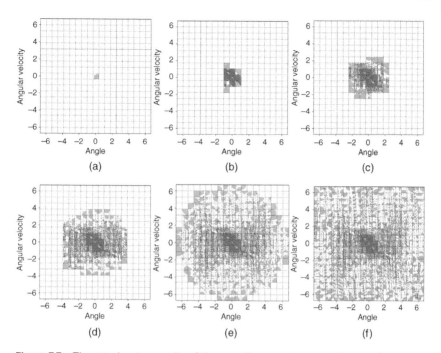

Figure 7.7 The step-by-step results of the sampling procedure and the sample gap obtained are provided that correspond to the results in Figures 7.5 and 7.6. It is evident that by acquiring more samples over different steps and expanding the learning area, the sample gaps are decreased effectively.

where skidding is possible. Therefore, Pepy et al. [2006] proposed a more realistic dynamical model of the vehicle which is implemented in this chapter (Figure 7.7).

According to Pepy et al. [2006], we present the dynamic model of the vehicle implemented. Let us define the states x and y as the coordinate of the center of gravity in the 2D space, θ as the orientation of the vehicle, v_y as the lateral velocity, and r as the rate of the orientation. Moreover, the input of the system is given by the front-wheel angle δ_f. Then, by assuming a constant longitudinal velocity v_x, the dynamical model of the vehicle can be written as follows:

$$\dot{v}_y = -\frac{C_{\alpha f}\cos\delta_f + C_{\alpha r}}{mv_x}v_y + \frac{-L_f C_{\alpha f}\cos\delta_f + L_r C_{\alpha r}}{I_z v_x}r + \frac{C_{\alpha f}\cos\delta_f}{m}\delta_f,$$

$$\dot{r} = \left(\frac{-L_f C_{\alpha f}\cos\delta_f + L_r C_{\alpha r}}{mv_x} - v_x\right)v_y$$

$$- \frac{L_f^2 C_{\alpha f}\cos\delta_f + L_r^2 C_{\alpha r}}{I_z v_x}r + \frac{L_f C_{\alpha f}\cos\delta_f}{I_z}\delta_f,$$

$$\dot{x} = v_x\cos\theta - v_y\sin\theta,$$

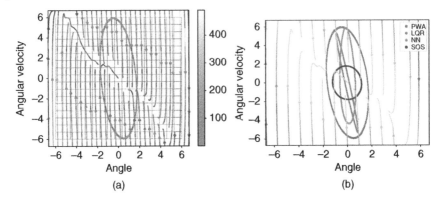

Figure 7.8 (a) The obtained ROA of the closed loop PWA system is illustrated for x_1 and $x_2 \in [-6, 6]$. The uniform grids denote the modes of the PWA system. Multiple trajectories of the system are shown in a phase portrait where color map represents the magnitude in the vector field. (b) The comparison results for ROA of the closed-loop system is illustrated for x_1 and $x_2 \in [-6, 6]$, together with the trajectories of the system. The comparison results for LQR, NN, and SOS are adapted from Chang et al. [2019].

$$\dot{y} = v_x \sin\theta + v_y \cos\theta,$$
$$\dot{\theta} = r, \tag{7.25}$$

where C_{xf}, and C_{xr} denote the cornering stiffness coefficients of the front and rear wheels. Moreover, the distance of the center of gravity from the front and rear wheels are given by L_f and L_r (Figure 7.8).

Identify and Control

Control objective is to minimize the distance of the vehicle from the goal point $(x, y)_{\text{goal}} = (70, 70)$ in the 2D map. To achieve the objective, we run the vehicle from some random initial position and yaw values. Then, identification and control procedures are done in a loop through different episodes. The longitudinal velocity of the vehicle is assumed to be constant in this system similar to Pepy et al. [2006]. Therefore, to minimize the cost given by the control objective, the vehicle converges to some circular path around the goal point, which is indeed the optimal path for the problem defined. Figure 7.5 contains the simulation results within an episode of learning, including the state and control signals, prediction error for each state, the value function, and the modes (Figure 7.9).

7.6.3 Comparison of Runtime Results

To analyze the computational aspects of the proposed technique, we provide the runtime results while learning the dynamics and obtaining the control for both

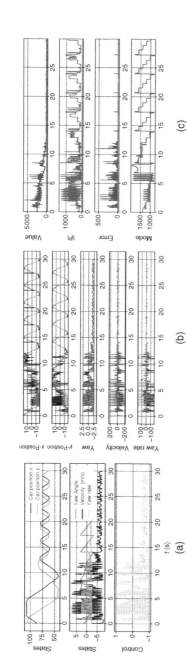

Figure 7.9 (a) The state and control signals are illustrated within an episode of learning. It can be clearly seen from the position signals that the vehicle is able to minimize the distance from the goal point and converge to a circular path around the goal point after some time of learning. (b) The graph denotes the evolutions of the value function, the norm of the control parameters for the active mode, the prediction error, and the active mode of the piecewise model that correspond to the results in Figure 7.9a. (c). Corresponding to Figure 7.9a, c, the prediction results of the learned model is compared with the original function learned is minimized. (c). Corresponding to Figure 7.9a. It can be seen that the value system within an episode of learning. It can be observed that the prediction signals shown by the black lines can match the ones obtained from the original dynamic.

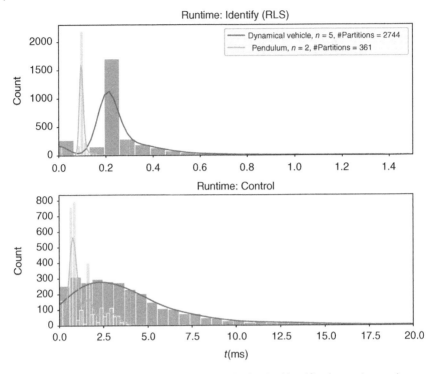

Figure 7.10 A comparison of the runtime results for the identification and control procedures separately is given for the implemented examples.

examples implemented. The proposed framework is considered as an online technique. Hence, in the applications, the computational complexity of the real-time identification and control becomes more important. Therefore, we here focus on the complexity of the online learning procedure rather than the verification technique which can be done offline. Figure 7.10 includes the runtime results separately for the identification and control units. Accordingly, the identifier and the controller can be updated in at most 1 and 20 ms, respectively. Accordingly, it can be observed that for the higher-dimensional system with also a larger number of partitions, the computations still remain in a tractable size that can allow real-time applications, considering the nature of the systems.

7.7 Summary

For regulating nonlinear systems with uncertain dynamics, a piecewise nonlinear affine framework was proposed in which each piece is responsible for learning and controlling over a partition of the domain locally. Then, in a particular case of the proposed framework, we focused on learning in the form of the well-known PWA systems, for which we presented an optimization-based verification approach that takes into account the estimated uncertainty bounds. We used the pendulum system as a benchmark example for the numerical results. Accordingly, a ROA resulting from the level set of a learned Lyapunov function is obtained. Furthermore, the comparison with other control approaches in the literature illustrates a considerable improvement in the ROA using the proposed framework. As another example, we implemented the presented approach on a dynamical vehicle system with considerably higher number of partitions and dimensions. The results demonstrated that the approach can scale efficiently and, hence, can be potentially implemented on more complex real-world problems in real time.

Bibliography

Dirk Aeyels and Joan Peuteman. A new asymptotic stability criterion for nonlinear time-variant differential equations. *IEEE Transactions on Automatic Control*, 43(7):968–971, 1998.

Amir Ali Ahmadi and Pablo A. Parrilo. Non-monotonic Lyapunov functions for stability of discrete time nonlinear and switched systems. In *Proceedings of the IEEE Conference on Decision and Control*, pages 614–621. IEEE, 2008.

Edoardo Amaldi, Stefano Coniglio, and Leonardo Taccari. Discrete optimization methods to fit piecewise affine models to data points. *Computers & Operations Research*, 75:214–230, 2016.

David S. Atkinson and Pravin M. Vaidya. A cutting plane algorithm for convex programming that uses analytic centers. *Mathematical Programming*, 69(1):1–43, 1995.

Ruxandra Bobiti and Mircea Lazar. A sampling approach to finding lyapunov functions for nonlinear discrete-time systems. In *Proceedings of the European Control Conference*, pages 561–566. IEEE, 2016.

Stephen Boyd and Lieven Vandenberghe. *Convex Optimization*. Cambridge University Press, 2004.

Stephen Boyd and Lieven Vandenberghe. Localization and cutting-plane methods. *Stanford EE 364b Lecture Notes*, 2007.

Valentina Breschi, Dario Piga, and Alberto Bemporad. Piecewise affine regression via recursive multiple least squares and multicategory discrimination. *Automatica*, 73:155–162, 2016.

Steven L. Brunton, Joshua L. Proctor, and J. Nathan Kutz. Discovering governing equations from data by sparse identification of nonlinear dynamical systems.

Proceedings of the National Academy of Sciences of the United States of America, 113(15):3932–3937, 2016.

Ya-Chien Chang, Nima Roohi, and Sicun Gao. Neural Lyapunov control. In *Proceedings of the International Conference on Neural Information Processing Systems*, pages 3245–3254, 2019.

Shaoru Chen, Mahyar Fazlyab, Manfred Morari, George J. Pappas, and Victor M. Preciado. Learning Lyapunov functions for piecewise affine systems with neural network controllers. *arXiv preprint arXiv:2008.06546*, 2020.

Steven Diamond and Stephen Boyd. CVXPY: A python-embedded modeling language for convex optimization. *The Journal of Machine Learning Research*, 17(1):2909–2913, 2016.

Jack Elzinga and Thomas G. Moore. A central cutting plane algorithm for the convex programming problem. *Mathematical Programming*, 8(1):134–145, 1975.

Milad Farsi, Yinan Li, Ye Yuan, and Jun Liu. A piecewise learning framework for control of nonlinear systems with stability guarantees. In *Proceedings of the Learning for Dynamics and Control Conference*. PMLR, 2022.

Giancarlo Ferrari-Trecate, Marco Muselli, Diego Liberati, and Manfred Morari. A clustering technique for the identification of piecewise affine systems. *Automatica*, 39(2):205–217, 2003.

Jean-Louis Goffin and Jean-Philippe Vial. On the computation of weighted analytic centers and dual ellipsoids with the projective algorithm. *Mathematical Programming*, 60(1):81–92, 1993.

Gurobi Optimization. Gurobi Optimizer Reference Manual, 2020.

Wassim M. Haddad and VijaySekhar Chellaboina. *Nonlinear Dynamical Systems and Control: A Lyapunov-Based Approach*. Princeton University Press, 2011.

Zhong-Ping Jiang and Yuan Wang. Input-to-state stability for discrete-time nonlinear systems. *Automatica*, 37(6):857–869, 2001.

Tobia Marcucci and Russ Tedrake. Mixed-integer formulations for optimal control of piecewise-affine systems. In *Proceedings of the ACM International Conference on Hybrid Systems: Computation and Control*, pages 230–239, 2019.

MOSEK ApS. The MOSEK optimization toolbox for Python manual, 2020.

Yu Nesterov. Complexity estimates of some cutting plane methods based on the analytic barrier. *Mathematical Programming*, 69(1):149–176, 1995.

Romain Pepy, Alain Lambert, and Hugues Mounier. Path planning using a dynamic vehicle model. In *2006 2nd International Conference on Information & Communication Technologies*, volume 1, pages 781–786. IEEE, 2006.

Steffen Rebennack and Vitaliy Krasko. Piecewise linear function fitting via mixed-integer linear programming. *INFORMS Journal on Computing*, 32(2):507–530, 2020.

Jie Sun, Kim-Chuan Toh, and Gongyun Zhao. An analytic center cutting plane method for semidefinite feasibility problems. *Mathematics of Operations Research*, 27(2):332–346, 2002.

Alejandro Toriello and Juan Pablo Vielma. Fitting piecewise linear continuous functions. *European Journal of Operational Research*, 219(1):86–95, 2012.

Ye Yuan, Xiuchuan Tang, Wei Zhou, Wei Pan, Xiuting Li, Hai-Tao Zhang, Han Ding, and Jorge Goncalves. Data driven discovery of cyber physical systems. *Nature Communications*, 10(1):1–9, 2019.

8

An Application to Solar Photovoltaic Systems

In this chapter, we present a case study on the design of optimal feedback control for a solar Photovoltaic (PV) system. The results presented in this chapter are published in Farsi and Liu [2019].

8.1 Introduction

Solar power as a renewable source of energy has attracted worldwide attention in recent years. Discussions regarding the maximization of solar energy accumulation have dominated research in this field, and many efforts have been made on the development of PV devices and their applications. Designing an effective control algorithm plays an important role in developing an efficient solar PV system. To this objective, various algorithms, known as Maximum Power Point Tracking (MPPT) methods, have already been presented in the literature of power electronics.

Among the conventional MPPT methods, perturb and observe (P&O) [Femia et al., 2005], incremental conductance algorithm [Lee et al., 2006, Sivakumar et al., 2015], and hill climbing algorithm (HC) [Xiao and Dunford, 2004] are the most favorable techniques. It has been shown that some of these approaches have advantages over others in terms of implementation complexity or performance. Furthermore, they are all easy to apply and, hence, become more suitable for low-cost applications. Comparative results can be found in Esram and Chapman [2007], Mohapatra et al. [2017], and Ram et al. [2017]. In spite of the simplicity of the conventional methods, they have shown a slow response to changes in the ambient temperature and solar radiation power [Ram et al., 2017]. Consequently, the deviation from the Maximum Power Point (MPP) of the system results in an extent of power loss which is proportional to the size of the implemented solar array. Hence, for relatively large solar arrays, in the trade-off between simplicity and performance, we tend to give priority to the latter since the amount of energy

Model-Based Reinforcement Learning: From Data to Continuous Actions with a Python-based Toolbox, First Edition. Milad Farsi and Jun Liu.

saved by the implementation of some elaborate techniques is appreciable enough to justify the extra cost brought.

The performance of MPPT methods can be analyzed by observing the operating point behavior in two phases: the convergence phase and the steady-state phase. In the convergence phase, the implemented control needs to be fast enough in its responses to immediately lead the operating point to the MPP of the system. This can save a considerable amount of energy when the system is exposed to a large-amplitude disturbance by the surrounding environment. On the other hand, in the steady-state phase, the operating point is constantly disturbed due to noises, model imperfections, and the structure of the controller circuit. Hence, the operating point needs to be continuously supervised by a high-performance control scheme to keep it within a desirable bound around the MPP. Although the oscillations caused by inefficient controllers usually take place in a small scale, they potentially waste a huge amount of energy in high-power implementations by failing to tightly track the ideal MPP, and also by draining considerable power on nonideal switching components.

Soft computing techniques, such as fuzzy logic control, artificial neural networks, and genetic algorithms, are effective tools in dealing with nonlinear problems. Hence, as a remedy of certain inherent drawbacks in conventional methods, numerous soft computing-based algorithms have been implemented on solar PV systems [Chiu and Ouyang, 2011, Tarek et al., 2013, Messai et al., 2011]. These approaches generally show improved results on the performance of MPPT control while each algorithm has its own constraints. For instance, despite that fuzzy logic is easy to implement and can provide a flexible design, it is widely accepted that designing fuzzy rules to satisfy a particular performance measure normally needs a considerable amount of knowledge and training; otherwise, using an inadequate number of membership functions will encourage oscillations around MPP [Ram et al., 2017].

From a control systems' point of view, solar PV systems can be modeled as a dynamical system for which more complicated and efficient control techniques can be exploited in comparison with the conventional MPPT methods. This provides many advantages in the design and analysis of solar PV systems since there already exists a vast literature of control systems that can be exploited to develop more efficient techniques (see, e.g. [Li et al., 2016, Bianconi et al., 2013, Chu and Chen, 2009]).

In Rezkallah et al. [2017], a Sliding Mode Control (SMC) approach with a saturation function has been presented that guarantees stability of the MPP of the system. In this approach, a design coefficient defining the sliding layer is chosen by trying different values and observing the convergence and steady-state results to find the optimum value. Although it illustrates improved results in the performance of the system in the nominal condition, the chattering problem still remains a major drawback since the saturation function and the controller gain chosen by trial and error cannot widely guarantee the performance for various settings of the solar PV system and the surrounding environment.

The double integral of the tracking error term can be used for constructing the sliding surface to eliminate steady-state error as well as to provide robust control responses against uncertainties [Tan et al., 2008]. While this contributes to the improved performance of the controller, it induces slow transient responses in the system. Hence, for tracking MPP of PV system, Pradhan and Subudhi [2016] develops an altered double integral SMC to speed up the convergence phase and alleviate the chattering effect. It stands to reason that, as an alternative method, the second-order SMC has advantages over the classical SMC methods in dealing with nonlinear systems. This justifies employing the second-order SMC in Sahraoui et al. [2016] by implementing a so-called "super-twisting algorithm." The approach is developed in Kchaou et al. [2017] for further moderating the chattering effect with the difference that, compared to Sahraoui et al. [2016], only one-loop control is used in Kchaou et al. [2017]. The simulation results presented in Kchaou et al. [2017] illustrate improved responses almost everywhere along the control signal. However, chattering effects can still be easily observed in the output power signal.

While there exist various approaches presented in the literature for improving the performance of the control in tracking the MPP, there is still no clear connection made between the configuration of the controller implemented and the performance obtained. Hence, performance analysis and defining the problem of maximizing the output power of PV systems in optimal control framework still remain as a challenging problem. To preserve a uniform quality in the performance of the system, a performance measure needs to be guaranteed.

In this chapter, we formulate and solve an optimal feedback control problem of solar PV systems that potentially bring many benefits in the applications. To obtain the optimal feedback control law, we consider a nonlinear affine model with a performance measure including a cross-weighting term. Hence, in contrast with the previous approaches, the performance analysis is additionally done by satisfying optimality conditions, which are adapted for a set of equilibrium points based on the incremental conductance approach. The obtained feedback controller, due to its suitable responses around the MPP, significantly decreases the undesired oscillations. Considering that the performance of the solar PV system is affected by the changing weather condition, we demonstrate the merits of the proposed controller under changing ambient temperature and solar radiation power.

The remaining of the chapter is organized as follows: in Section 8.2, we will investigate the model details and parameters involved in the solar PV system and the boost converter considered in this chapter. Furthermore, we will see how they are paired together to let the control scheme regulate the operating point of the

Table 8.1 Nomenclature.

Symbol	Description
I_{ph}	Light-generated current
I_s	Reverse saturation current
n	Ideality factor of PV cell
V_T	Thermal voltage
N_p, N_s	Parallel and series branches in PV module
R_{sh}, R_s	Shunt and series resistance of PV module
V_{oc}, I_{sc}	Open-circuit voltage and short-circuit current
L, C	Inductor and capacitor of DC-DC converter
R_L	Parasitic resistance of the inductor
R_0	Applied resistive load
P_a, V_a, i_a	Power, voltage, and current of solar array
v_C	Output voltage of DC-DC converter
V_o	Desired output voltage of DC-DC converter
k_{PID}	A vector of PID parameters

system as required. Sections 8.3 and 8.4 present the main results of this chapter. In Section 8.3 we first introduce an optimal control problem to minimize the deviation from the MPP of the system. Then, we show that the optimal control law exists with respect to the defined cost functional. Moreover, by modifying the formulated optimal control problem, we derive an optimal voltage control for solar PV systems. Section 8.4 addresses two main challenges in controlling real-world solar PV systems. In the Section 8.41, we propose an algorithm to realize the obtained control law without a complete knowledge of the system parameters and only by samples obtained from the output voltage and current of the PV system. In the Section 8.4.2, we discuss the considerations for implementing the approach under nonuniform insolation. In Section 8.5, the simulation results are provided for both model-based and model-free approaches, under uniform and nonuniform insolation.

A list of parameters used is given in Table 8.1.

8.2 Problem Statement

This section formulates the control system and provides sufficient details on the model for later use in the controller design procedure. As shown in Figure 8.1, in

Figure 8.1 DC-DC boost converter used to interface the load to the solar array.

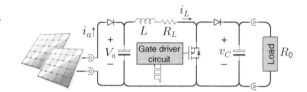

order to control the operating point of the system, a DC-DC boost converter can be used to couple the PV array with the load. The following two subsections present the model for describing the PV array and the boost converter, respectively.

8.2.1 PV Array Model

A solar PV array contains a number of PV modules. Suppose all the PV modules used in the array have identical electrical characteristics. Likewise, it is assumed that the surrounding environment affecting the solar panels, such as the ambient temperature, power density of the solar radiation, and wind speed, do not change considerably from one module to another.

Figure 8.2 illustrates the equivalent circuit used to model an array of the PV modules arranged in N_p parallel branches, where each branch contains N_s modules in series. Applying Kirchhoff's current law at the top node results in

$$i_a = N_p I_{ph} - N_p I_s \left(\exp\left(\frac{\frac{R_s i_a}{N_p} + \frac{V_a}{N_s}}{nV_T} \right) - 1 \right) - I_{sh}, \qquad (8.1)$$

wherein I_{ph}, I_s, n, and V_T are the light-generated current, reverse saturation current, ideality factor, and thermal voltage of the PV module, respectively. Moreover, considering the equivalent shunt resistor R_{sh} and the series resistor R_s of the PV module in the circuit, we can write the current flowing through the shunt resistor as follows:

$$I_{sh} = \frac{R_s i_a + \frac{N_p}{N_s} V_a}{R_{sh}}. \qquad (8.2)$$

Figure 8.2 The equivalent electrical model of the solar array.

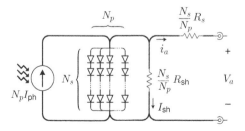

Hence, the output voltage V_a of the PV array can be calculated in terms of the output current i_a as follows:

$$V_a = N_s n V_T \ln\left(\frac{N_p I_{\text{ph}} + N_p I_s - i_a - I_{\text{sh}}}{N_p I_s}\right) - \frac{N_s}{N_p} R_s i_a. \tag{8.3}$$

Then the first-order and the second-order derivative of PV array voltage with respect to the output current can be computed as follows:

$$\frac{\partial V_a}{\partial i_a} = \frac{-N_s n V_T}{N_p I_{\text{ph}} + N_p I_s - i_a - I_{\text{sh}}} - \frac{N_s}{N_p} R_s, \tag{8.4}$$

$$\frac{\partial^2 V_a}{\partial i_a^2} = \frac{-N_s n V_T}{(N_p I_{\text{ph}} + N_p I_s - i_a - I_{\text{sh}})^2}.$$

Note that, since the output power of the PV array is less sensitive to the changes in R_{sh}, the dependence of I_{sh} to V_a is disregarded to simplify the computations of the derivatives in (8.4). More details on the model of the PV array are provided in Tsai [2010].

8.2.2 DC-D C Boost Converter

Dynamical systems with a finite number of subsystems are generally known as switched systems. In such systems, a switching strategy orchestrates the switching action among these subsystems to ensure stability and performance. Switched affine systems introduce an important class of switched systems with a constant input that brings much convenience to design and application.

Consider the following model of the switched system:

$$\dot{\mathbf{x}} = A_\sigma \mathbf{x} + B_\sigma V_a, \tag{8.5}$$

$$\mathbf{y} = C_\sigma \mathbf{x},$$

where $A_\sigma \in \mathbb{R}^{n\times n}, B_\sigma \in \mathbb{R}^n$ and $C_\sigma^T \in \mathbb{R}^n$ denote the system matrices. The active subsystem is given by the piecewise-constant signal $\sigma : [0, \infty) \to \{0,1\}$. By choosing the current of the inductor i_L and the voltage of the capacitor v_C as the states of the system, the state vector becomes $\mathbf{x} = [i_L \quad v_C]^T$. The input V_a plays the role of the power source for the system (8.5). As seen in the model of the PV array (8.3), this input depends on the state since i_a equals i_L for the coupled DC-DC converter and the solar PV array. Moreover, V_a is affected by the changes in the light-generated current I_{ph} and operating temperature T, considering that V_T is related to the temperature in (8.3). However, since the input irradiance and temperature are slowly changing parameters compared to the switched current flowing through the inductor, V_a is considered as a function of the state only. Henceforth, we regard $V_a : D \to \Xi$, where $D \subseteq \mathbb{R}^2_+$ and $\Xi \subseteq \mathbb{R}_+$ are domains of interest with $\mathbb{R}^n_+$ denoting the n-dimensional positive-real space.

In this chapter, a boost converter is used to formulate the problem, while a similar approach can be applied to different configurations of converters. Details on the design and switched system model of DC-DC converters can be found in Noori et al. [2016], Deaecto et al. [2010], and Noori et al. [2014]. For a typical boost converter, the switched affine model can be constructed by the following system matrices:

$$A_0 = \begin{bmatrix} -R_L/L & 0 \\ 0 & -1/R_0C \end{bmatrix} , \quad A_1 = \begin{bmatrix} -R_L/L & -1/L \\ 1/C & -1/R_0C \end{bmatrix} ,$$

$$B_1 = B_0 = \begin{bmatrix} 1/L \\ 0 \end{bmatrix} , \quad C_1 = C_0 = \begin{bmatrix} 1 & 1 \end{bmatrix} .$$

The load applied to the system is denoted by $R_0 \in \Omega \subseteq \mathbb{R}_+$. Moreover, L and C are positive constants that denote the values of the inductor and the capacitor, respectively. In this model, the inductor is supposed to be nonideal, for which a parasitic resistance, indicated by R_L, is considered in series. Likewise, the leakage current of the output capacitor can be modeled as a resistor in parallel combination with the output load; however, it is disregarded in this model since the structure of the system matrices is not affected by that.

To obtain the pulse-width modulation (PWM)-controlled model of the system, the switched affine system is overapproximated by the convex hull of the subsystems

$$\dot{\mathbf{x}} = (1 - u(t))(A_0\mathbf{x} + B_0 V_a) + u(t)(A_1\mathbf{x} + B_1 V_a),$$

where $u(t) : [0, \infty) \to [0,1]$ gives the duty-cycle values. This results in the average model of the system as follows:

$$\dot{\mathbf{x}} = u(t)g(\mathbf{x}) + f(\mathbf{x}, V_a, R_0), \tag{8.6}$$

with $\mathbf{x} = [i_L \; v_C]^T$ and

$$g(\mathbf{x}) = \begin{bmatrix} -\dfrac{1}{L}v_C \\ \dfrac{1}{C}i_L \end{bmatrix} , \quad f(\mathbf{x}, V_a, R_0) = \begin{bmatrix} -\dfrac{R_L}{L}i_L + \dfrac{1}{L}V_a \\ -\dfrac{1}{R_0C}v_C \end{bmatrix} ,$$

where $\mathbf{x} := \mathbf{x}(t) \in D \subseteq \mathbb{R}_+^2$ for $t \in [0, \infty)$ and $\mathbf{x}(0) = x_0$. Moreover, $f : D \times \Xi \times \Omega \to \mathbb{R}^2$ and $g : D \to \mathbb{R}^2$ are vector-valued functions. It should be noted that the problem is formulated for inverted PWM generators, where, for instance, the sampled control $u(t^k) = 0$ at kth time step will keep the transistor in ON mode within $t \in [t^k, t^{k+1})$ for $k \in \mathbb{N}$. This can be simply adapted for noninverted PWM devices by again inverting the duty-cycle value generated by the obtained control signal.

Furthermore, by the following substitution:

$$u(t) = \frac{V_a - R_L i_L}{v_C} + \omega_c(t), \tag{8.7}$$

the system dynamics can be rewritten as follows:

$$\frac{d}{dt}\begin{bmatrix} i_L \\ v_C \end{bmatrix} = \omega_c(t)\underbrace{\begin{bmatrix} -\dfrac{1}{L}v_C \\ \dfrac{1}{C}i_L \end{bmatrix}}_{g(x)} + \underbrace{\begin{bmatrix} 0 \\ \dfrac{V_a i_L - R_L i_L^2}{v_C C} - \dfrac{v_C}{R_0 C} \end{bmatrix}}_{f(x, V_a, R_0)}, \tag{8.8}$$

where the control $\omega_c(t) \in W \subseteq \mathbb{R}$ for $t \in [0, \infty)$ is chosen from the set of all admissible controls Γ. The first part of control (8.7) can be seen as an equivalent control that requires $\frac{di_L}{dt} = 0$ for (8.6). Furthermore, the simple linear dependence of $\frac{di_L}{dt}$ on the control input $\omega_c(t)$ in (8) facilitates the stability analysis done later in the main results.

After modeling the switched system using a nonlinear affine system, in the next step, we exploit an inverse optimal control approach (see, e.g. [Moylan and Anderson, 1973, Bernstein, 1993, Haddad and Chellaboina, 2011, Haddad and L'Afflitto, 2016]) to pose and solve an optimal control problem for tracking the MPP of the solar PV system.

8.3   Optimal Control of PV Array

In this section, we consider a particular form of the cost functional to regulate the performance of the solar PV system. We first present conditions needed for optimality and stability of (8.8) with respect to a set of equilibrium points and a given performance measure. By formulating an optimal MPPT problem, we then confirm that the stability and optimality conditions obtained indeed hold for the obtained control law.

Lemma 8.1   *Consider the system (8.8) with the cost functional*

$$J(x_0, \omega_c(\cdot)) = \lim_{T \to \infty} \int_0^T L\left(\xi\left(x, V_a(x)\right), \omega_c(t)\right) \, dt, \tag{8.9}$$

where x is the solution starting from $x_0 \in D$, and $L : \mathbb{R} \times W \to \mathbb{R}$ is the running cost. Moreover, $\xi : D \times \Xi \to \mathbb{R}$ defines the equilibrium set as follows:

$$E = \{x \in D : \xi(x, V_a(x)) = 0\}. \tag{8.10}$$

Suppose there exists a C^1 function $V : D \times \Xi \to \mathbb{R}$ and a control law $\omega_c^* = \phi(x, V_a(x))$ with $\phi : D \to W$ such that

$$V(x, V_a(x)) = 0 \qquad \text{for} \quad x \in E, \tag{8.11}$$

$$V(x, V_a(x)) > 0 \qquad \text{for} \quad x \in D, x \notin E, \tag{8.12}$$

$$\phi(x, V_a(x)) = 0 \qquad \text{for} \quad x \in E, \tag{8.13}$$

$$\begin{aligned} V_x^T[\phi(x, V_a(x))g(x) + f(x, V_a(x), R_0)] < 0 \\ \text{for} \quad x \in D, \quad x \notin E, \quad R_0 \in \Omega, \end{aligned} \tag{8.14}$$

$$H\left(x, V_a(x), \phi(x, V_a(x))\right) = 0 \quad \text{for} \qquad x \in D, \quad R_0 \in \Omega, \tag{8.15}$$

$$H\left(x, V_a(x), \omega_c\right) \geq 0 \qquad \text{for} \quad x \in D, \quad \omega_c \in W, \quad R_0 \in \Omega, \tag{8.16}$$

where V_x and H denote, respectively, the partial derivatives with respect to the state as follows:

$$V_x := \frac{\partial V}{\partial x} + \frac{\partial V}{\partial V_a}\frac{\partial V_a}{\partial x},$$

and the Hamiltonian defined by

$$\begin{aligned} H\left(x, v, \omega\right) &= \mathbf{L}\left(\xi(x, v), \omega\right) \\ &+ V_x^T\left(\omega g(x) + f(x, v, R_0)\right). \end{aligned}$$

Then, with the feedback control rule, the solutions of (8.8) converge to the set E. Moreover, the feedback control rule minimizes the performance functional in the sense that

$$J(x_0, \omega_c^*(\cdot)) = \min_{\omega_c \in \Gamma} J(x_0, \omega_c(\cdot)), \tag{8.17}$$

where

$$J\left(x_0, \omega_c^*(\cdot)\right) = V(x_0, V_a(x_0)). \tag{8.18}$$

Proof: Conditions (8.11) to (8.14) guarantee attractivity of the set E since $\mathbf{V}$ is a Lyapunov function of the system (8.8).

The derivative of the Lyapunov function is given by

$$\dot{\mathbf{V}}(\mathbf{x}, V_a(\mathbf{x})) = V_x^T\left(\omega_c g(\mathbf{x}) + f(\mathbf{x}, V_a(\mathbf{x}), R_0)\right), \tag{8.19}$$

then we add the running cost to both sides of (8.19) to obtain

$$\begin{aligned} \mathbf{L}\left(\xi(\mathbf{x}, V_a(\mathbf{x})), \omega_c(t)\right) &= \mathbf{L}\left(\xi(\mathbf{x}, V_a(\mathbf{x})), \omega_c(t)\right) - \\ &\dot{\mathbf{V}}(\mathbf{x}, V_a(\mathbf{x})) + V_x^T\left(\omega_c(t)g(\mathbf{x}) + f(\mathbf{x}, V_a(\mathbf{x}), R_0)\right). \end{aligned} \tag{8.20}$$

By integrating both sides from 0 to T and letting $T \to \infty$, we obtain

$$\begin{aligned} J(x_0, \omega_c(\cdot)) &= \lim_{T \to \infty} \int_0^T [\mathbf{L}\left(\xi(\mathbf{x}, V_a(\mathbf{x})), \omega_c(t)\right) - \dot{\mathbf{V}}(\mathbf{x}, V_a(\mathbf{x})) \\ &+ V_x^T\left(\omega_c(t)g(\mathbf{x}) + f(\mathbf{x}, V_a(\mathbf{x}), R_0)\right)]dt \\ &= \lim_{T \to \infty} \int_0^T [-\dot{\mathbf{V}}(\mathbf{x}, V_a(\mathbf{x})) \\ &+ H\left(\mathbf{x}, V_a(\mathbf{x}), \omega_c(t)\right)]dt \end{aligned}$$

$$= \mathbf{V}(x_0, V_a(x_0)) - \lim_{T \to \infty} \mathbf{V}\left(x_T, V_a(x_T)\right)$$

$$+ \lim_{T \to \infty} \int_0^T H\left(\mathbf{x}, V_a(\mathbf{x}), \omega_c(t)\right) dt$$

$$\geq \mathbf{V}(x_0, V_a(x_0)), \tag{8.21}$$

where this concludes (8.17) by defining the Hamiltonian and using (8.16) and (8.18). ∎

8.3.1 Maximum Power Point Tracking Control

The goal of MPPT techniques is to maximize the output power of the PV array which is measured as follows:

$$P_a = V_a i_a.$$

The output power is obviously a function of the state and the state-dependent input of the switched system. Looking at (8.3), the input V_a is only related to the first state, where by considering the average model, we have $i_a = i_L$. Hence, according to the incremental conductance approach [Lee et al., 2006, Sivakumar et al., 2015], the stationary point of the output power with respect to the inductor current, i.e. points such that

$$\frac{\partial P_a}{\partial i_L} = \frac{\partial V_a}{\partial i_L} i_L + V_a = 0, \tag{8.22}$$

defines a set of equilibrium points that addresses the MPP of the solar array. By this knowledge, one can observe that choosing (8.9) in the following form:

$$J(x_0, \omega_c(\cdot))$$

$$= \lim_{T \to \infty} \int_0^T [\mathbf{L}_1(\xi) + \omega_c(t)\mathbf{L}_2(\xi) + S\omega_c(t)^2]dt, \tag{8.23}$$

along with the appropriate choices of functions $\mathbf{L}_1, \mathbf{L}_2 : \mathbb{R} \to \mathbb{R}$ penalize the deviation of the operating point of PV array from $\xi = \frac{\partial P_a}{\partial i_L} = 0$, which guarantees the MPP, and simultaneously regulates the control input. In (8.23), $\mathbf{L}_1$ and $\mathbf{L}_2$ can be written in terms of x and V_a by using (8.22), and S is a positive constant given by design considerations.

As declared in Haddad and Chellaboina [2011], it is evident that involving the cross-weighting term, with $\mathbf{L}_2 \neq 0$, not only brings extra flexibility into the design but also illustrates better transient performance in terms of peak overshoot over the case $\mathbf{L}_2 = 0$.

Remark 8.1 The uniform irradiance only leads to one maximum point in the power–voltage (P–V) characteristic curves that can be located with minimizing the defined cost functional, and it is known to be the global MPP of the system. However, under nonuniform irradiance, only reaching a local MPP is guaranteed. To overcome the resulting nonconvexity, more actions are needed that will be discussed in detail later in Section 8.4.

Having defined the cost functional, we need to solve the minimization problem (8.17) to achieve the control law, which optimally leads the operating point of the system to MPP.

Theorem 8.1 *Consider the nonlinear affine dynamical system (8.8) and performance measure (8.23) with functions* $L_1, L_2 : \mathbb{R} \to \mathbb{R}$ *chosen, respectively, as follows:*

$$L_1(\xi) = \frac{p^2}{4S}(\bar{\xi} - 1)^2 \xi^2, \quad L_2(\xi) = p\xi, \quad p > 0, \tag{8.24}$$

where

$$\xi = \frac{\partial P_a}{\partial i_L}, \quad \bar{\xi} = \left(\frac{\partial^2 V_a}{\partial i_L^2} i_L + 2\frac{\partial V_a}{\partial i_L}\right)\left(\frac{v_C}{L}\right). \tag{8.25}$$

Then, by the choice of the inductor L with $0 < R_L < \lim_{i_L \to 0}\left\{|\frac{\partial V_a}{\partial i_L}|\right\}$ *and the feedback control rule (8.7) constructed by*

$$\omega_c^* = \frac{1}{2S}p\xi(\bar{\xi} - 1), \tag{8.26}$$

the solutions of the system (8.8) converge to the set E defined by (8.10) with (8.25), and the performance measure (8.23) is minimized, for a given positive constant S.

Proof: As stated, the MPP is given by the stationary point of the output power with respect to the state i_L as in (8.22). Thus, for tracking the MPP, we need to guarantee that the set E defined by choosing $\xi = \frac{\partial P_a}{\partial i_L}$ is attractive. Therefore, regarding the structure of the running cost, the optimal value function is supposed to possess the following form:

$$V = \frac{1}{2}p\left(\frac{\partial P_a}{\partial i_L}\right)^2 = \frac{1}{2}p\left(\frac{\partial V_a}{\partial i_L}i_L + V_a\right)^2. \tag{8.27}$$

According to the HJB equation and using the optimal value function defined in (8.27),

$$0 = \inf_{\omega_c(\cdot) \in \Gamma} \{H(x, V_a, \omega_c) \tag{8.28}$$

$$= L(\xi, \omega_c) + V_x^T(\omega_c g(x) + f(x, V_a, R_0))\},$$

holds along the optimal path given by the optimal control law chosen from the set of all admissible controls Γ, where $L(\xi, w_c)$ is the running cost chosen as in (8.23).

With regard to (8.8) and (8.27), the optimal value function is independent of v_C. Also, v_C does not appear in (8.3) and (8.5). Therefore, **V** only changes along i_L, i.e.

$$
\mathbf{V}_x = \begin{bmatrix} \dfrac{\partial \mathbf{V}}{\partial i_L} \\[2mm] \dfrac{\partial \mathbf{V}}{\partial v_C} \end{bmatrix} = \begin{bmatrix} \dfrac{\partial \mathbf{V}}{\partial i_L} + \dfrac{\partial \mathbf{V}}{\partial V_a}\dfrac{\partial V_a}{\partial i_L} + \dfrac{\partial \mathbf{V}}{\partial\left(\dfrac{\partial V_a}{\partial i_L}\right)}\dfrac{\partial}{\partial i_L}\left(\dfrac{\partial V_a}{\partial i_L}\right) \\[4mm] 0 \end{bmatrix}
\tag{8.29}
$$

$$
= \begin{bmatrix} p\left(\dfrac{\partial V_a}{\partial i_L}i_L + V_a\right)\left(\dfrac{\partial^2 V_a}{\partial i_L^2}i_L + 2\dfrac{\partial V_a}{\partial i_L}\right) \\[3mm] 0 \end{bmatrix}.
$$

Now, by substitution of the optimal value and the running cost, the Hamiltonian is obtained as follows:

$$
H(x, V_a, \omega_c) = \mathbf{L}_1(\xi) + \omega_c\mathbf{L}_2(\xi) + S\omega_c^2
\tag{8.30}
$$

$$
+ \mathbf{V}_x^T(w_c g(x) + f(x, V_a, R_0))
$$

$$
= \mathbf{L}_1(\xi) + \omega_c\mathbf{L}_2(\xi) + S\omega_c^2 + \dfrac{\partial V}{\partial i_L}\left(\dfrac{-v_C\omega_c}{L}\right)
$$

$$
+ \dfrac{\partial V}{\partial v_C}\left(\dfrac{1}{C}i_L\omega_c + \dfrac{V_a i_L - R_L i_L^2}{v_C C} - \dfrac{v_C}{R_0 C}\right)
$$

$$
= \mathbf{L}_1(\xi) + \omega_c\mathbf{L}_2(\xi) + S\omega_c^2
$$

$$
+ p\left(\dfrac{\partial V_a}{\partial i_L}i_L + V_a\right)\left(\dfrac{\partial^2 V_a}{\partial i_L^2}i_L + 2\dfrac{\partial V_a}{\partial i_L}\right)\left(\dfrac{-v_C\omega_c}{L}\right).
$$

Then, the optimal control law minimizing the given Hamiltonian is obtained by solving the following equation:

$$
\dfrac{\partial H(x, V_a, \omega_c)}{\partial \omega_c} = \mathbf{L}_2(\xi) + 2\omega_c S
$$

$$
+ p\left(\dfrac{\partial V_a}{\partial i_L}\,i_L + V_a\right)\left(\dfrac{\partial^2 V_a}{\partial i_L^2}i_L + 2\dfrac{\partial V_a}{\partial i_L}\right)\left(\dfrac{-v_C}{L}\right) = 0.
$$

As a result, the optimal control law is given by

$$
\omega_c^* = \dfrac{-1}{2S}\left[\mathbf{L}_2(\xi)\right.
\tag{8.31}
$$

$$
\left. + p\left(\dfrac{\partial V_a}{\partial i_L}i_L + V_a\right)\left(\dfrac{\partial^2 V_a}{\partial i_L^2}i_L + 2\dfrac{\partial V_a}{\partial i_L}\right)\left(\dfrac{-v_C}{L}\right)\right].
$$

To shorten the computations, an intermediary function $\bar{\xi}$ is defined as (8.25), by which Eqs. (8.31) and (8.31) become

$$
H(x, V_a, \omega_c) = \mathbf{L}_1(\xi) + \omega_c(\mathbf{L}_2(\xi) - p\xi\bar{\xi}) + S\omega_c^2,
$$

$$
\omega_c^* = \dfrac{-1}{2S}(\mathbf{L}_2(\xi) - p\xi\bar{\xi}),
\tag{8.32}
$$

respectively. Furthermore, the substitution of ω_c^* in the Hamilton–Jacobi–Bellman (HJB) equation yields

$$\inf_{\omega_c(\cdot) \in \Gamma} \left\{ H(x, V_a, \omega_c) \right\} = H(x, V_a, \omega_c^*) \tag{8.33}$$

$$= \mathbf{L}_1(\xi) - \frac{1}{2S}(\mathbf{L}_2(\xi) - p\xi\bar{\xi})^2$$

$$+ \frac{1}{4S}(\mathbf{L}_2(\xi) - p\xi\bar{\xi})^2$$

$$= \mathbf{L}_1(\xi) - \frac{1}{4S}(\mathbf{L}_2(\xi) - p\xi\bar{\xi})^2.$$

To determine the optimal control input, it only remains to choose $\mathbf{L}_1$ and $\mathbf{L}_2$ functions such that the optimality and stability conditions are satisfied as required in Lemma 8.1. Regarding the structure of the value function in (8.27), conditions (8.11) and (8.12) hold. Furthermore, to guarantee the asymptotic stability, we need to verify (8.14) for the obtained feedback control. Using (8.8) and (8.29), $\dot{V}$ is obtained as the following:

$$\dot{V} = \mathbf{V}_x^T(w_c g(x) + f(x, V_a, R_0))$$

$$= \left[p\left(\frac{\partial V_a}{\partial i_L}i_L + V_a \right)\left(\frac{\partial^2 V_a}{\partial i_L^2}i_L + 2\frac{\partial V_a}{\partial i_L} \right) \right]^T$$
$$\times \left(\omega_c \begin{bmatrix} -\frac{1}{L}v_C \\ \frac{1}{C}i_L \end{bmatrix} + \begin{bmatrix} 0 \\ \frac{V_a i_L - R_L i_L^2}{v_C C} - \frac{v_C}{R_0 C} \end{bmatrix} \right)$$

$$= p\left(\frac{\partial V_a}{\partial i_L}i_L + V_a \right)\left(\frac{\partial^2 V_a}{\partial i_L^2}i_L + 2\frac{\partial V_a}{\partial i_L} \right)\left(\frac{-v_C \omega_c}{L} \right). \tag{8.34}$$

With $\bar{\xi}$ defined in (8.25) and the optimal control law (8.32), this results in

$$\dot{V}|_{\omega_c^*} = -p\xi\xi\left(\frac{1}{2S}(\mathbf{L}_2(\xi) - p\xi\bar{\xi}) \right)$$

$$= \frac{1}{2S}p\xi\bar{\xi}\mathbf{L}_2(\xi) - \frac{1}{2S}(p\xi\bar{\xi})^2, \tag{8.35}$$

wherein p and S are positive constants. Moreover, according to (8.25), the sign of $\bar{\xi}$ depends on $\frac{\partial V_a}{\partial i_L}$ and $\frac{\partial^2 V_a}{\partial i_L^2}$ which both are known to be negative values, considering electrical characteristics of solar PV cells. See, for instance, the current-voltage characteristic curves of a solar array in Figures 8.3 and 8.4, wherein the voltage is strictly decreasing and concave-downward with respect to the current. As a result, $\bar{\xi}$ only takes negative values. In (8.35), the second term is obviously negative definite; however, to decide the sign of $\dot{V}$, we still need to inspect the values taken by

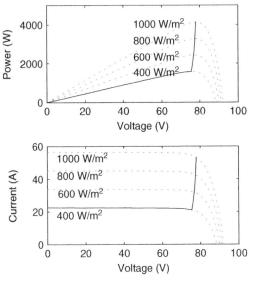

Figure 8.3 Output results of the simulated solar array by changing the irradiance power from 400 to 1000 W/m², where the solid lines represent the track of the MPP obtained by using the proposed NOC approach.

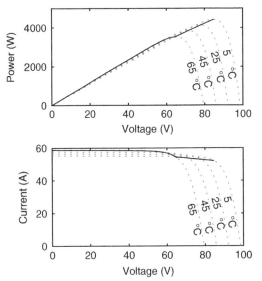

Figure 8.4 Output results of the simulated solar array by changing the ambient temperature from 5 to 65 °C, where the solid lines represent the track of the MPP by using the proposed NOC.

ξ and $\mathbf{L}_2$ that can be any positive or negative real values. To deal with this undetermined situation, let $\mathbf{L}_2$ as in (8.24), then,

$$\dot{\mathbf{V}}|_{\omega_c^*} = \frac{1}{2S}(p\xi)^2\bar{\xi} - \frac{1}{2S}(p\xi\bar{\xi})^2 < 0$$

is obtained that is clearly negative definite with $\bar{\xi} < 0$ for $x \notin E$, and satisfies (8.14). Moreover, this completes the optimal control input as in (8.26) by which (8.13) is also verified for the set E defined in (8.10).

Now, we choose $\mathbf{L}_1$ as in (8.24) to satisfy the HJB equation. With the substitution of $\mathbf{L}_1$ and $\mathbf{L}_2$, it follows from (8.33) that

$$
\begin{aligned}
H(x, V_a, \omega_c^*) &= \frac{p^2}{4S}\xi^2(\bar{\xi} - 1)^2 + \frac{1}{2S}p\xi(\bar{\xi} - 1)(p\xi - p\xi\bar{\xi}) \\
&\quad + S\left(\frac{1}{2S}p\xi(\bar{\xi} - 1)\right)^2 \\
&= \frac{p^2}{4S}\xi^2(\bar{\xi} - 1)^2 - \frac{p^2}{2S}\xi^2(\bar{\xi} - 1)^2 + \frac{p^2}{4S}\xi^2(\bar{\xi} - 1)^2 \\
&= 0.
\end{aligned}
\tag{8.36}
$$

Thus, the HJB condition of optimality holds as in (8.15). In the last step, we need to inspect the Hamiltonian given by (8.32) to assure positivity for all admissible control $\omega_c \in W$, as required in (8.16),

$$
\begin{aligned}
H(x, V_a, \omega_c) &= \mathbf{L}_1(\xi) + \omega_c(\mathbf{L}_2(\xi) - p\xi\bar{\xi}) + S\omega_c^2 \\
&= \frac{p^2}{4S}\xi^2(\bar{\xi} - 1)^2 + \omega_c(p\xi - p\xi\bar{\xi}) + S\omega_c^2 \\
&= \left(\frac{p}{2\sqrt{S}}\xi(\bar{\xi} - 1)\right)^2 - \omega_c p\xi(\bar{\xi} - 1) + (\sqrt{S}\omega_c)^2 \\
&= \left(\frac{p}{2\sqrt{S}}\xi(\bar{\xi} - 1) - \sqrt{S}\omega_c\right)^2 \geq 0.
\end{aligned}
\tag{8.37}
$$

This completes the conditions needed to be verified in Lemma 8.1. Hence, the obtained feedback control also satisfies optimality conditions and regulates the performance of MPPT controller with the performance measure given by (8.23).

Furthermore, the equilibrium currents of the closed-loop system (8.8), with control law (8.26), can be obtained by using the first state equation of the system (8.8):

$$
i_L^{eq} =
\begin{cases}
-V_a \big/ \dfrac{\partial V_a}{\partial i_L}, \\[2ex]
\left(\dfrac{L}{v_C} - 2\dfrac{\partial V_a}{\partial i_L}\right)\left(1 \big/ \dfrac{\partial^2 V_a}{\partial i_L^2}\right),
\end{cases}
\tag{8.38}
$$

where the latter is obviously negative and out of D, by negativity of the partial derivatives. Hence, only the first equilibrium current belongs to D. Plugging in the valid equilibrium current in the second-state equation of the system (8.8) yields a relation for the equilibrium voltage of the capacitor as follows:

$$v_C^{eq} = \pm \frac{\sqrt{-R_0 \left(R_L + \frac{\partial V_a}{\partial i_L}\right)}}{\frac{\partial V_a}{\partial i_L}} V_a. \tag{8.39}$$

Similarly, one of the equilibrium voltage relations is always negative and out of D. Thus, for a fixed R_0, the operating point of the system converges to the only valid equilibrium point

$$(i_L^{eq}, v_C^{eq}) = \left(-V_a \Big/ \frac{\partial V_a}{\partial i_L}, -\frac{\sqrt{-R_0 \left(R_L + \frac{\partial V_a}{\partial i_L}\right)}}{\frac{\partial V_a}{\partial i_L}} V_a\right), \tag{8.40}$$

which takes values in E for different $R_0 \in \Omega$.

Accordingly, to achieve a positive real equilibrium voltage, we also need, the design specification of the inductor satisfy $0 < R_L < |\frac{\partial V_a}{\partial i_L}|$, for any $x \in D$. Hence, a practical and safe choice of the upper bound for R_L is $\inf_{x \in D} \left\{|\frac{\partial V_a}{\partial i_L}|\right\} = \lim_{i_L \to 0} \left\{|\frac{\partial V_a}{\partial i_L}|\right\}$, which can be estimated from the voltage-current characteristic curve or experiment results of the solar PV array near open-circuit state. ∎

Remark 8.2 Considering relation (8.40), the equilibrium current does not directly depend on the output load. Hence, the proposed control law can independently follow the MPP regardless of the applied load, while the equilibrium voltage can be chosen by only regulating R_0. In the solar PV array connected to the AC grid, the load is usually controlled by a separate PI controller to regulate the output voltage v_C at a fixed level which is vital for correct power injection to the grid. Moreover, to ensure the valid operation of the boost converter, we need $v_C > V_a$. This suggests a lower bound to the applied load as follows:

$$R_0 > -\left(\frac{\partial V_a}{\partial i_L}\right)^2 \Big/ \left(R_L + \frac{\partial V_a}{\partial i_L}\right), \tag{8.41}$$

for any $x \in D$, which can be used to make an estimation of Ω.

8.3.2 Reference Voltage Tracking Control

In Section 8.3.1, we proposed a framework to optimally control the solar PV array to gain the maximum power. As a second application of the proposed framework, we design an optimal feedback control rule to regulate the output voltage of the solar PV array to a reference value.

Corollary 8.1 *Consider the nonlinear affine dynamical system (8.8) and performance measure (8.23) with* $\mathbf{L}_1$ *and* $\mathbf{L}_2$ *chosen respectively as (8.24), where*

$$\xi = V_a - V_{\text{ref}}, \quad \bar{\xi} = \frac{\partial V_a}{\partial i_L} \frac{v_C}{L}. \tag{8.42}$$

Then, by the feedback control rule (8.26) constructed by (8.42), the solutions of the system (8.8) converge to the set E defined by (8.10), and the performance measure (8.23) is minimized.

Proof: Regarding the structure of the running cost, the optimal value function is supposed to possess a quadratic form as follows:

$$\mathbf{V} = \frac{1}{2}p(V_a - V_{\text{ref}})^2. \tag{8.43}$$

By taking the derivative, we obtain

$$\dot{\mathbf{V}} = \mathbf{V}_x^T(w_c g(x) + f(x, V_a, R_0))$$

$$= \left[\begin{array}{c} p(V_a - V_{\text{ref}}) \left(\dfrac{\partial V_a}{\partial i_L} \right) \\ 0 \end{array} \right]^T \left(\omega_c \left[\begin{array}{c} -\dfrac{1}{L} v_C \\ \dfrac{1}{C} i_L \end{array} \right] + \left[\begin{array}{cc} 0 \\ V_a i_L - R_L i_L^2 & \dfrac{v_C}{v_C C} - \dfrac{v_C}{R_0 C} \end{array} \right] \right)$$

$$= p(V_a - V_{\text{ref}}) \left(\frac{\partial V_a}{\partial i_L} \right) \left(\frac{-v_C \omega_c}{L} \right). \tag{8.44}$$

For the closed-loop system, by substituting the control rule (8.26) constructed by (8.42), the following can be concluded

$$\dot{\mathbf{V}} = \underbrace{\frac{p^2}{2S}(V_a - V_{\text{ref}})^2}_{(+)} \underbrace{\left(\frac{-\partial V_a}{\partial i_L} \frac{v_C}{L} \right)}_{(-)} \left(\frac{\partial V_a}{\partial i_L} \frac{v_C}{L} - 1 \right) < 0.$$

This is obtained by the fact that the first partial derivative is always negative that also makes the last term negative, while the other terms multiplied are all positive. Hence, the set E defined by choosing ξ as in (8.42) is attractive. The rest is followed in the similar way as the proof of Theorem 8.1 by defining Hamiltonian as (8.30) with the choices of ξ and $\bar{\xi}$ as in (8.42), by which, in addition, conditions (8.15) and (8.16) of the Lemma 8.1 are satisfied. ∎

While the control rule obtained can be employed to regulate the output voltage of the solar PV array in various applications, in Section 8.4 and later in the simulation results, we will see how it is particularly useful in partial shading condition.

Remark 8.3 It should be noted that the control objective is to control the output voltage of the PV array, which is different than the goal of MPPT controller obtained in the last subsection. Hence, both parameters p and S may be independently adjusted for each case.

Remark 8.4 Similar to Kchaou et al. [2017] and Rezkallah et al. [2017], the proposed approach can be considered as a model-based control approach since the partial derivatives appeared in the control rule depend on the model parameters as in (8.4). In the following section, we introduce a procedure, according to the obtained optimal scheme, to develop a model-free control by approximating the partial derivatives. In addition, we combine the results of Theorem 8.1 and Corollary 8.1 to deal with the partial shading phenomenon as a well-known imperfection in the operation of real-world PV arrays.

8.3.3 Piecewise Learning Control

Consider the following system,

$$
\frac{d}{dt}\begin{bmatrix} i_L \\ v_C \\ \xi \end{bmatrix} = \begin{bmatrix} -\dfrac{1}{L}v_C u(t) - \dfrac{R_L}{L}i_L + \dfrac{1}{L}V_a \\ \dfrac{1}{C}i_L u(t) - \dfrac{1}{R_0 C}v_C \\ \dfrac{d}{dt}\left(\dfrac{\partial V_a}{\partial i_L}i_L + V_a \right) \end{bmatrix}, \tag{8.45}
$$

which is constructed by the average model of the solar PV system (8.6), augmented with ξ as the third state. It should be noted that the states i_L and v_C can be directly measured. In addition, since we can measure V_a, then ξ can be approximated. The only concern is obtaining the partial derivative term that can be also obtained by measurements done. This will be discussed in detail in the next section.

Now, this nonlinear system contains the DC-DC converter's dynamics, the PV array model through function V_a, and ξ which can be used to reach the MPP. Let us assume that we do not have access to the parameters of this system that is indeed a common scenario in real-world applications. Hence, we need to approximate such system by only our observations of $\begin{bmatrix} i_L & v_C & \xi \end{bmatrix}^T$. The problem formulated fits best in the piecewise learning framework presented in Chapter 7. Accordingly, we use the following model to approximate this unknown system:

$$
\dot{x} = W_\sigma \Phi(x) + \sum_{j=1}^{m} W_{j\sigma}\Phi(x)u_j + d_\sigma, \tag{8.46}
$$

where W_σ and $W_{j\sigma} \in \mathbb{R}^{n \times p}$ are the matrices of the coefficients for $\sigma \in \{1, 2, \dots, n_\sigma\}$ and $j \in \{1, 2, \dots, m\}$, with a set of differentiable bases $\Phi(x) = [\phi_1(x) \quad \dots \quad \phi_p(x)]^T$, and n_σ denoting the total number of pieces.

Considering that the MPP is given by $\xi = 0$, to set up a learning MPPT controller, it only remains to choose Q in the control objective (5.2) in a way that the third component of the state is penalized.

8.4 Application Considerations

In this section, we address two main challenges of establishing MPPT control in real-world applications.

8.4.1 Partial Derivative Approximation Procedure

In the optimal approach proposed in Section 8.4, it is supposed that the exact values of partial derivatives, given by (8.4), are available at any $t \in \mathbb{R}_+$. However, in the real-world implementation of PV arrays, there exist numerous parameters affecting the output power characteristics of the PV array that cannot be directly measured or estimated. In the literature, some effort has been made on online parameter identification of solar PV arrays, while application of the MPPT methods using only the output voltage and current measurements of the PV array is often preferred. This is because of their simplicity and robustness while using no extra knowledge of the surrounding environment and electrical characteristics of the solar cell, which makes them less expensive for applications as well.

To set up the presented optimal control approach, we only need the partial derivatives in (8.4), which can be obtained approximately by using the sampled output voltage and output current of the PV array. Consider the output current i_a, the light-generated current I_{ph}, and the ambient temperature T as three major parameters affecting the output voltage of the PV array, given in (8.3). Compared to the output current, the solar irradiation and ambient temperature are changing slowly. Hence, it is assumed $|di_a| \gg |dI_{ph}|$ and $|di_a| \gg |dT|$. Then the rate of changes of the output voltage of the array is approximated for sufficiently small $|dt|$ as follows:

$$\frac{dV_a}{dt} \simeq \frac{\partial V_a}{\partial i_a} \frac{di_a}{dt},$$

and this yields an estimation of the partial derivative as follows:

$$\frac{\partial V_a}{\partial i_a} \simeq \frac{dV_a}{dt} \bigg/ \frac{di_a}{dt}. \tag{8.47}$$

Remark 8.5 This can be considered as a strong assumption applied to the problem that might adversely affect the performance of the system when the rate of changes in the solar irradiation and the ambient temperature is relatively high.

However, since in the realistic weather condition, the irradiation and temperature inputs will eventually reach a stable condition with a tolerable rate of changes, the controller will also be able to survive from sudden disturbances and retrieve the track of the MPP in a short time.

Moreover, the second-order derivative can be similarly written as following:

$$
\frac{d^2 V_a}{dt^2} = \frac{d}{dt}\left(\frac{dV_u}{dt}\right)
$$

$$
\simeq \frac{\partial}{\partial i_a}\left(\frac{\partial V_a}{\partial i_a}\frac{di_a}{dt}\right)\frac{di_a}{dt}
$$

$$
= \left(\frac{\partial^2 V_a}{\partial i_a^2}\frac{di_a}{dt} + \frac{\partial V_a}{\partial i_a}\frac{\partial}{\partial i_a}\left(\frac{di_a}{dt}\right)\right)\frac{di_a}{dt}
$$

$$
= \frac{\partial^2 V_a}{\partial i_a^2}\left(\frac{di_a}{dt}\right)^2. \tag{8.48}
$$

This is obtained using the fact that the derivative of the output current is independent of the output current, according to the first state equation of the system (8.6). Then the second-order partial derivative becomes

$$
\frac{\partial^2 V_a}{\partial i_a^2} \simeq \frac{d^2 V_a}{dt^2}\bigg/\left(\frac{di_a}{dt}\right)^2, \tag{8.49}
$$

as $dt \to 0$. For the implementation of the proposed approach, the values given by (8.47) and (8.49) are needed to be calculated at each time sample. Thus, for a sufficiently small sampling time, $\tau = t^k - t^{k-1}$, (8.47), and (8.49) can be measured at $t = t^k$, /respectively, as follows:

$$
\frac{\partial V_a}{\partial i_a}\bigg|_{t=t^k} \simeq \left(\frac{V_a^k - V_a^{k-1}}{\tau}\right)\bigg/\left(\frac{i_a^k - i_a^{k-1}}{\tau}\right) = \frac{\Delta V_a^k}{\Delta i_a^k},
$$

$$
\frac{\partial^2 V_a}{\partial i_a^2}\bigg|_{t=t^k} \simeq \left(\frac{\Delta^2 V_a^k}{\tau^2}\right)\bigg/\left(\frac{\Delta i_a^k}{\tau}\right)^2 = \frac{\Delta V_a^k - \Delta V_a^{k-1}}{\left(\Delta i_a^k\right)^2}, \tag{8.50}
$$

where

$$
\Delta i_a^k = i_a^k - i_a^{k-1}, \tag{8.51}
$$

$$
\Delta V_a^k = V_a^k - V_a^{k-1},
$$

$$
\Delta V_a^k - \Delta V_a^{k-1} = V_a^k - 2V_a^{k-1} + V_a^{k-2}.
$$

Hence, while the division by Δi_a is allowed, the relations in (8.50) approximate the partial derivatives required for constructing the feedback control. It is worth noting that the signals obtained in (8.51) may be prone to high-frequency noises in applications. However, considering the smooth properties of (8.4), we can safely employ low-pass filters as long as they do not induce slow responses.

In the convergence phase, since the inductor current is strictly increasing or decreasing toward its steady-state value within a period of time, the chance to meet $\Delta i_a \to 0$ is sufficiently low. Therefore, the approximations done by the divisions in relations (8.50) are expected to be valid until the steady state is achieved. In other words, the feedback controller supplied by these approximations will be able to converge to the MPP.

Furthermore, the performance of the controller highly depends on the steady-state responses as well, where oscillations of i_L around its steady-state value result in some stationary points of i_L that make the denominator $\Delta i_a \to 0$. Let $\kappa > 0$ be the minimum value allows dividing by, that is chosen by design considerations. Then, based on the second derivative in (8.51), $|\Delta i_a|$ can take at least $\sqrt{\kappa}$ to yield a valid division. Hence, relations (8.50) are used as long as Δi_a is greater than $\sqrt{\kappa}$. Otherwise, the samples are accumulated by $\Sigma_{\Delta i}$, $\Sigma_{\Delta V}$, and $\Sigma_{\Delta^2 V}$, without any update done in the approximation values. Afterward, once the condition $|\Sigma_{\Delta i}| > \sqrt{\kappa}$ is met, the divisions are done and the approximation values are updated by the accumulated samples.

As stated, the technique established for the approximation of the partial derivatives relies on the magnitude of changes in i_L. If the controller chooses to let this current rest for a while, there will be no longer updates on the approximation values as well. Thus, it gives the operating point the opportunity to diverge from MPP as far as possible. In other words, a minimum perturbation is always required on i_L to perform a valid approximation and to decide a suitable control input as quickly as possible. Therefore, when the amplitude of changes on i_L do not satisfy the condition $|\Delta i_a| > \sqrt{\kappa}$, a constant positive or negative value, in accordance with the sign of changes on i_L, is added to the control input to encourage the perturbations. This procedure is illustrated in Algorithm 8.1, where the approximated values are only updated when the condition is satisfied; otherwise, the samples are accumulated for future use, and a small perturbation is continuously added to the PWM value in each iteration.

In the simulation results, the proposed algorithm will be implemented to control a sample solar PV array under uniform and nonuniform insolation.

8.4.2 Partial Shading Effect

Partial shading phenomenon is widely studied in the literature as one of the factors resulting in current-voltage characteristic curve mismatching among PV modules. Although nonuniform insolation caused by partially shaded PV modules is known as the most likely scenario, there exist other possible imperfections, such as the production tolerance, accumulated dust, and aging [Patel and Agarwal, 2008, Spertino and Akilimali, 2009], that can promote the mismatching effect.

Algorithm 8.1 An algorithm for the approximation of partial derivatives.

1: **procedure** APPROXIMATION OF PARTIAL DERIVATIVES
   Input:
2: Samples obtained from i_a and V_a;
   Output:
3: $\quad \dfrac{\partial V_a}{\partial i_a}\Big|_{t^k}, \dfrac{\partial^2 V_a}{\partial i_a^2}\Big|_{t^k}$;
   Initialization:
4: Choose $\varepsilon, \kappa > 0$;
5: Set $\Sigma_{\Delta i}, \Sigma_{\Delta V}, \Sigma_{\Delta^2 V} = 0$;
6: $\quad$ **while** (*true*) **do**
7: $\qquad$ Read samples i_a^k, V_a^k;
8: $\qquad$ Update $\Delta i_a^k, \Delta V_a^k$ and $\Delta^2 V_a^k$ by (8.51);
9: $\qquad$ **if** $|\Delta i_a| > \sqrt{\kappa}$ **then**
10: $\qquad\qquad$ Update $\dfrac{\partial V_a}{\partial i_a}\Big|_{t^k}, \dfrac{\partial^2 V_a}{\partial i_a^2}\Big|_{t^k}$ using (8.50);
11: $\qquad$ **else**
12: $\qquad\qquad \Sigma_{\Delta i} := \Sigma_{\Delta i} + \Delta i_a^k$;
13: $\qquad\qquad \Sigma_{\Delta V} := \Sigma_{\Delta V} + \Delta V_a^k$;
14: $\qquad\qquad \Sigma_{\Delta^2 V} := \Sigma_{\Delta^2 V} + \Delta^2 V_a^k$;
15: $\qquad\qquad$ **if** $|\Sigma_{\Delta i}| > \sqrt{\kappa}$ **then**
16: $\qquad\qquad\qquad$ Update:
17: $\qquad\qquad\qquad \dfrac{\partial V_a}{\partial i_a}\Big|_{t^k} \simeq \dfrac{\Sigma_{\Delta V}}{\Sigma_{\Delta i}}$;
18: $\qquad\qquad\qquad \dfrac{\partial^2 V_a}{\partial i_a^2}\Big|_{t^k} \simeq \dfrac{\Sigma_{\Delta^2 V}}{(\Sigma_{\Delta i})^2}$;
19: $\qquad\qquad\qquad$ Reset $\Sigma_{\Delta i}, \Sigma_{\Delta V}, \Sigma_{\Delta^2 V} = 0$;
20: $\qquad\qquad$ **else**
21: $\qquad\qquad\qquad$ Add perturbation:
22: $\qquad\qquad\qquad \omega_c := \omega_c + \varepsilon \text{Sign}(\Sigma_{\Delta i})$;
23: $\qquad\qquad$ **end if**
24: $\qquad$ **end if**
25: $\quad$ **end while**
26: **end procedure**

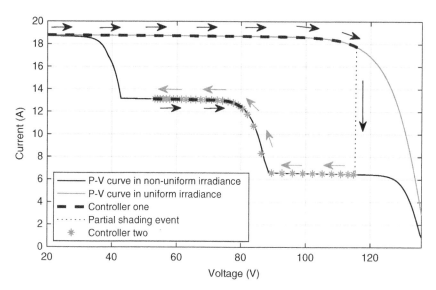

Figure 8.5 Evolutions of the operating point of system on $I-V$ curve before and after partial shading event, where controllers are defined by (8.52).

As shown in Figures 8.5 and 8.6, the mismatching in current–voltage ($I-V$) curves caused by partial shading effect leads to some local maxima in the $P-V$ characteristic curve of the solar array. Consequently, conventional MPPT methods, such as P&O, incremental conductance, and HC algorithms, as well as control system approaches, such as SMC, second-order SMC, and double integrator control, possibly fail in tracking the global maximum since they search locally by following the direction that increases the output power. Hence, a higher-level control is required to systematically [Bidram et al., 2012] or randomly [Sundareswaran et al., 2014] switch among the local areas to search for the greatest MPP of the solar array. In this regard, some approaches are already presented in the literature, such as power increment technique [Koutroulis and Blaabjerg, 2012], load-line MPPT [Kobayashi et al., 2003, Ji et al., 2011], and instantaneous operating power optimization approach [Carannante et al., 2009], that exploit one of the conventional methods at some stage to identify the local MPP corresponding to the current area of interest (for more details see [Bidram et al., 2012]). Therefore, to boost the performance of these algorithms, the control rule suggested by Theorem 8.1 can be implemented as an alternative to the conventional methods and previously presented control system approaches.

The proposed optimal control framework can be combined by algorithms presented in the literature to tackle the mismatching effect appeared in nonuniform

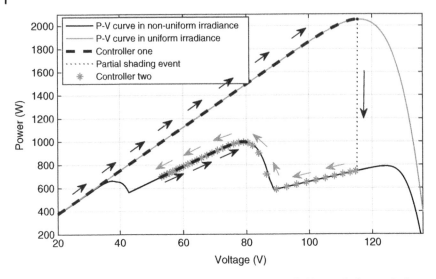

Figure 8.6 Evolutions of the operating point of system on P-V curve before and after partial shading event, where controllers are defined by (8.52).

insolation. In this chapter, we exploit the load-line technique together with the optimal voltage control obtained in Corollary 8.1 to relocate the operating point after the partial shading event. In contrast with [Ji et al., 2011], to implement [Kobayashi et al., 2003], additional circuits are required to measure V_{oc} and I_{sc} online. Hence, we compose the main controller by the following control rules:

Controller_One : (8.7) with (8.26) using (8.25),

Controller_Two : (8.7) with (8.26) using (8.42),   (8.52)

where the controller choice is governed by Algorithm 8.2 that exploits the load-line technique in Ji et al. [2011]. A graphical representation of the approach together with the simulation results will be provided in Section 8.5.

8.5 Simulation Results

To assess the proposed approach under a realistic condition, the Canadian Solar CS6X-335M-FG module [Canadian Solar, 2022] has been simulated in MATLAB/Simulink® with the obtained control rule to generate the maximum power in different weather conditions. This module contains 72 solar cells, and the electrical characteristics were listed in Table 8.2. In this simulation, the PV array is composed by 12 modules placed in 6 parallel branches where there exist 2 modules in series at each branch. As illustrated in Figure 8.1, a DC-DC boost

Table 8.2 Electrical data of the CS6X-335M-FG module [Canadian Solar, 2022].

Under STC (1000 W/m² and 25 °C)	Value
Nominal max. power (P_{max})	335 W
Open-circuit voltage (V_{oc})	46.1 V
Short-circuit current (I_{sc})	9.41 A
Temperature coefficient (P_{max})	−0.41%/°C
Temperature coefficient (V_{oc})	−0.31%/°C
Temperature coefficient (I_{sc})	0.053%/°C
Nominal module operating temperature (NMOT)	43 ± 2 °C

converter is used to regulate the load applied to the PV array that provides the control over the operating point of the solar array.

Control input (8.7) is constructed by the optimal control law obtained in (8.26). According to the cost functional (8.23) defined by (8.24), increasing S penalizes the control effort. Moreover, by looking at (8.26), one can observe that only the proportion of S and p appeared in the control rule. Hence, by arbitrarily fixing p, which scales the value function, and by relatively changing S, the control parameters corresponding to a desirable performance can be obtained. Regarding Remark 8.3, for each of Controllers (8.52), the parameters are chosen as $[S, p] = [5.88, \ 1 \times 10^{-5}]$ and $[S, p] = [1, \ 1 \times 10^{-2}]$, respectively.

Since the average model of the converter was considered, the continuous value given by the control law needs to be converted back into a quantized signal that complies with the number of subsystems. Therefore, any value given by $u(t) \in [0,1]$ is considered as a duty cycle to constantly generate a pulsed signal to drive the switch within a particular period of time which depends on the frequency chosen for PWM. Moreover, the output load is controlled by a PI controller to regulate the output voltage v_C to a fixed DC voltage level V_o. For PV arrays connected to AC grid, this can be replaced by a DC-AC inverter. A schematic diagram of the system is given in Figure 8.7.

In this simulation, the parameters of the boost converter are set to have the following values: $L = 0.2$ mH, $R_L = 1$ Ω, and $C = 2500$ μF. Also, switching components are chosen such that for the diode, we have $V_{On} = 0.6$ V, on resistance= 0.3 Ω, and off conductance= 10^{-8} Ω^{-1}, and for the switch, on resistance = 10^{-2} Ω, and off conductance= 10^{-6} Ω^{-1}. The PWM frequency is set to be 65 kHz, where the duty cycle value is updated by the controller every 0.1 ms. Moreover, we set

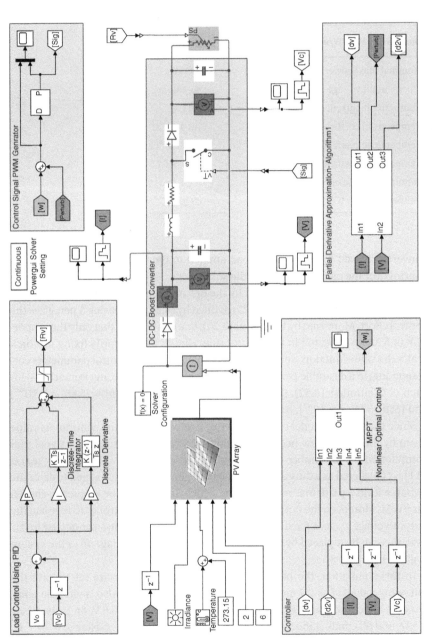

Figure 8.7 A sketch of the simulated solar PV system together with the proposed control approach in Matlab Simulink.

Algorithm 8.2 The algorithm of the control scheme in partial shading condition.

1: **procedure** CONTROL SCHEME IN PARTIAL SHADING
  Input:
2: Samples obtained from i_a and V_a;
  Output: Subcontroller chosen from (8.52).
3: **Initialization:**
4: Choose $\varsigma > 0$;
5: Define $V_{ref} = 0$, *Controller* = Controller_One;
6:   **while** (*true*) **do**
7:     Read samples i_a^k, V_a^k;
8:     **if** (partial_shading_condition) **then**
9:       $V_{ref} = \frac{N_s V_{oc}}{N_p I_{sc}} i_a^k$;
10:       *Controller* = Controller_Two;
11:       **while** ($|V_a^k - V_{ref}| > \varsigma$) **do**
12:         Read sample V_a^k;
13:       **end while**
14:       *Controller* = Controller_One;
15:     **end if**
16:   **end while**
17: **end procedure**

$k_{PID} = [6, 10, 0]$. The desired output voltage level is assumed to be $V_0 = 120\,$V for any part of the simulations, unless explicitly mentioned otherwise.

8.5.1 Model and Control Verification

Since the function of solar PV systems is mainly affected by the changes in the input irradiance and temperature, the performance of the proposed nonlinear optimal control (NOC) approach is investigated by changing these parameters. Figure 8.3 denotes the output results of the simulated solar array by changing the irradiance power from 400 to 1000 W/m². Similarly, the next set of graphs in Figure 8.4 is obtained by changing the ambient temperature from 5 to 65 °C. In both figures, the system starts up from the rest situation, i.e. $x = 0$. Then, the operating point continues moving on the $I-V$ and $P-V$ characteristic curves determined by the input irradiance and temperature, until it reaches its MPP. The system operates around the MPP, while the input does not change. Once the input starts moving to the next value on a ramp, the controller also regulates the operating point to the corresponding MPP of the system. Hence, it is evident in these figures that the proposed controller can successfully lead the operating point to the MPP and track it in the presence of a disturbance applied on the temperature and irradiance.

8.5.2 Comparative Results

In the comparison results of the system, the obtained control law is compared with two recent approaches targeting on the performance improvement of the generated control signal: SMC [Rezkallah et al., 2017] and second-order SMC [Kchaou et al., 2017]. For a fair comparison, the boost converter and solar PV system were chosen to remain the same for all simulated approaches.

As the first scenario, we generated a random fast-changing signal as the input irradiance of the system while the operating temperature is fixed at 25 °C. This signal is shown in Figure 8.8d. By running the simulation with these inputs, the output power is captured for the optimal control and the other two approaches in

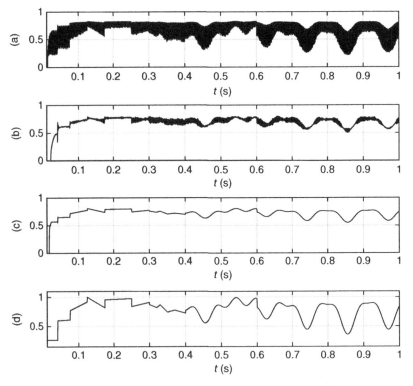

Figure 8.8 The obtained control signal (NOC) compared to SMC and second-order SMC under the changing irradiance shown in (d). (a) SMC control signal. (b) Second-order SMC control signal. (c) Proposed NOC control signal. (d) Irradiance (kW/m^2).

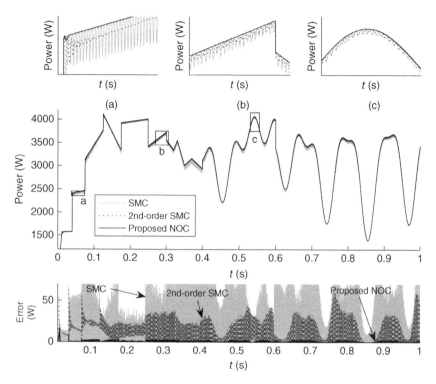

Figure 8.9 Comparison results of the output power under the changing irradiance shown in Figure 8.8d, where some parts of the graph are magnified in subfigures (a–c). The error graph denotes the comparative error for the proposed NOC compared to SMC and second-order SMC (see the text for details).

Figure 8.9. In the second scenario, the same was done by fixing the input irradiance at 800 W/m^2 and applying a changing temperature, as shown in Figure 8.10d.

By the output power comparison results, in Figures 8.9 and 8.11, we can observe the SMC control reaches to the MPP faster than the other approaches, while it shows a constant chattering effect almost everywhere. The second-order SMC shows improved results on the chattering effect; however, it fails to demonstrate a robust performance within the simulation. See, for instance, Figures 8.9a and 8.11b, wherein the second-order SMC manages to efficiently decrease the chattering effect, whereas, in Figures 8.9c and 8.11c, the amplitude of oscillations around MPP is almost as large as the SMC control. In addition, the second-order SMC illustrates slow responses under fast changes of the input, i.e. in the convergence phase, and hence loses the track of MPP in fast-changing weather conditions. For instance, see Figures 8.9a and 8.11a, respectively. In contrast with the second-order SMC control, the obtained control with a guaranteed cost is able

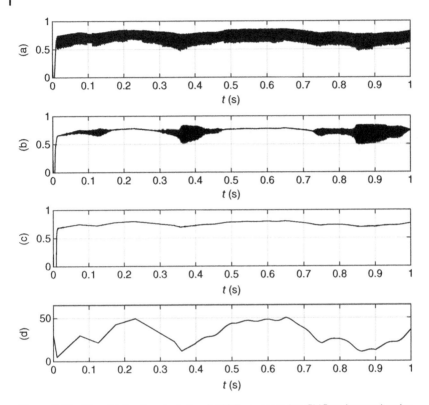

Figure 8.10 The obtained control signal (NOC) compared to SMC and second-order SMC under the changing ambient temperature shown in (d). (a) SMC control signal. (b) Second-order SMC control signal. (c) Proposed NOC control signal. (d) Temperature (°C).

to maintain a constant performance for different weather conditions. Moreover, from the comparative error results given in Figures 8.9 and 8.11, it can be clearly seen that the proposed control law outperforms the other two methods, where the error graph, for any control approach, is obtained by the absolute difference of the corresponding output power with the maximum of all three approaches at any time instant $t \in [0, 1]$. This is also evident in the corresponding control signals given in Figures 8.8 and 8.10, that the optimal control signal illustrates the smoothest response with almost no chattering effect.

8.5.3 Model-Free Approach Results

To verify the model-free controller using the proposed approximation procedure, Algorithm 8.1 was implemented with a changing irradiance as the input. As mentioned, κ is a sufficiently small value that yields a valid division. This

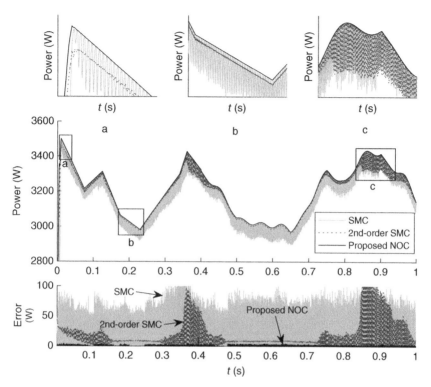

Figure 8.11 Comparison results of the output power under the changing ambient temperature shown in Figure 8.10d, where some parts of the graph are magnified in subfigures (a–c). The error graph denotes the comparative error for the proposed NOC compared to SMC and second-order SMC (see the text for details).

can be chosen by starting from large values and decreasing until a minimum amplitude of oscillations is observed in the steady state. Moreover, ε is a small perturbation recursively added to control that can affect output signal of the PWM generator within a few time steps. Hence, it is chosen with respect to the PWM resolution of the device. A possible choice is a fraction of $1/PWM\_resolution$. In this simulation, the parameters are chosen as $\kappa = 1 \times 10^{-3}$ and $\varepsilon = 5 \times 10^{-3}$. Figure 8.12 shows the approximated partial derivatives, perturbation, and output power signals corresponding to the input irradiance. It is evident by Figure 8.12e that NOC operating based on the approximation of partial derivatives are able to tightly track the ideal maximum power curve. Although, as expected, the output power corresponding to the model-free controller is not as smooth as the results obtained by the model-based controller, it does still illustrate satisfactory results.

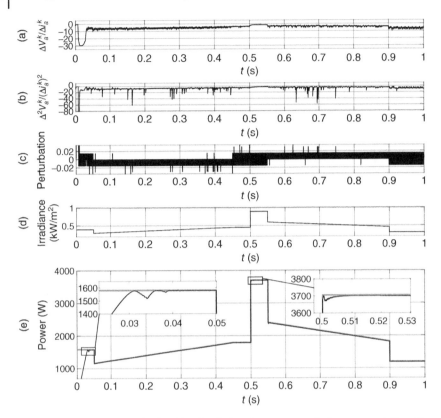

Figure 8.12 The results obtained by simulating the system with the proposed Algorithm 3.2, which illustrate, respectively: (a) and (b) First- and second-order partial derivatives. (c) Perturbation signal added to improve the estimation. (d) The variable input irradiance applied to the solar PV array. (e) The output power.

8.5.4 Piecewise Learning Results

In this section, we implement the piecewise learning controller on a sample PV system. To do so, we use a PV system as a black box for which the characterizing P–V and I–V graphs are given in Figure 8.13 together with the MPP. Targeting the system (8.45), we choose the observation vector as $\begin{bmatrix} i_L & v_c & \xi \end{bmatrix}$. To start learning, we need to define the partitions of the piecewise model. Therefore, we uniformly grid the state space with the number of points given by the vector $[11, 11, 11]$, which corresponds to the state vector. We choose only the linear and constant bases that yields a linear Piecewise Affine (PWA) model. Regarding the discussion on approximating the partial derivatives, and the convergence of the piecewise

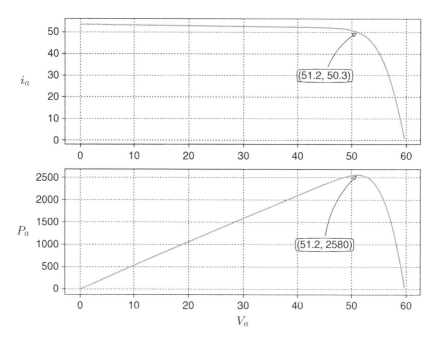

Figure 8.13 *I–V* and *P–V* graphs are shown that characterize the solar PV system used for piecewise learning control as an example. Moreover, the MPP is denoted.

model learner, the system needs to be persistently stimulated with some input signal. Hence, a sinusoidal probing signal with the amplitude 0.03 and frequency 76 Hz is added to the control. Moreover, we define the objective function with $Q = \text{diag}([0, 0, 10^2])$ and $R = [10^2]$.

According to the objective defined, it is expected that after enough time of learning, the controller drives ξ to zero. According to the definition of the MPP this guarantees operating in the MPP of the system. It can be observed from Figure 8.14 that ξ indeed converges to zero after about 0.4 seconds of training. Hence, the learning control can achieve the MPPT objective. This is also shown by the PV array voltage V_a that can track the MPP voltage given by Figure 8.13.

8.5.5 Partial Shading Results

Considering the partial shading effect as shown in Figure 8.15, we run a simulation of the PV array in both uniform and nonuniform insolation. In this regard, we implemented Algorithm 8.2 with $\varsigma = 1$ and the dc link voltage is set to be $V_o = 160$. Furthermore, we used Algorithm 8.1 to estimate the partial derivatives.

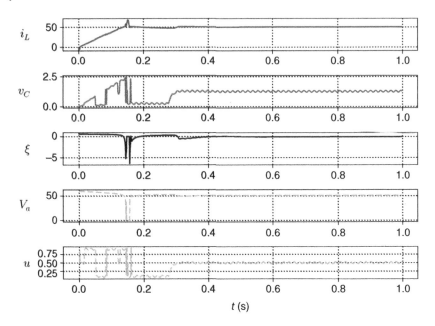

Figure 8.14 The learning result of the solar PV system, given by Figure 8.13, are shown. It can be observed that after 0.4 seconds, ξ converges to zero, that guarantees operating in the MPP. It is also evident by the V_a signal that the PV array voltage can track the MPP voltage 51.2 V given by Figure 8.13.

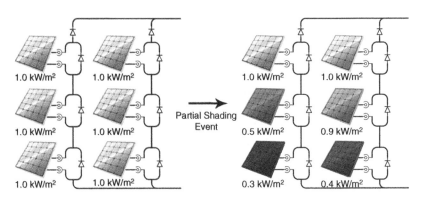

Figure 8.15 The solar PV array illustrating the partial shading condition considered in the simulation results.

Evolutions of the operating point on I–V and P–V curves are given in Figures 8.5 and 8.6, respectively. The system is run from zero initial state assuming uniform insolation. As seen in P–V curve of the system Figure 8.6, Controller One (8.52) is used to reach the MPP of the system. Once the system is exposed to the nonuniform insolation, some local maxima appear in the P–V curve. Consequently, the partial shading condition is detected by the algorithm and the reference voltage is calculated as $V_{ref} = 49.5$ V. This voltage is tracked by Controller Two (8.52) until V_a is ς-close to the reference voltage, where ς can be chosen with respect to the open-circuit voltage of the PV array so that remaining in the neighborhood of the global MPP is assured. Once the output voltage is arrived in the neighborhood of the calculated V_{ref}, Controller One (8.52) is again activated to track the MPP in that local area which is expected to be the global MPP. In Figure 8.16, output power, voltage, and current are illustrated that correspond to Figures 8.5 and 8.6. Furthermore, Figure 8.16 denotes the control signal and switchings between the subcontrollers given by (8.52).

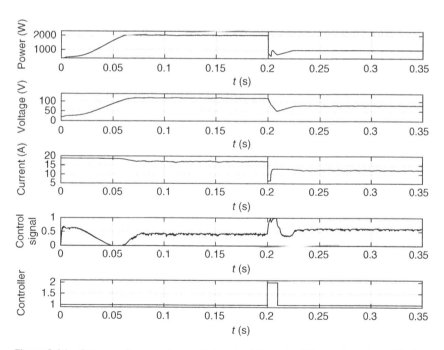

Figure 8.16 Output voltage and current signals of the solar PV array together with the control signal, where the shading event, corresponding to Figures 9 and 10, is detected at 0.2 seconds.

8.6 Summary

Motivated by the lack of performance analysis of the solar PV system in the literature, we developed an optimal feedback control approach to improve the convergence and steady-state responses of the system. The nonlinear nonquadratic cost as a performance measure, which involves the cross-weighting term, introduced a degree of freedom in the stability and optimality analysis. Then, we obtained a nonlinear optimal feedback controller by minimizing the corresponding Hamiltonian. The idea is exploited to establish two controllers for tracking the MPP and a given reference voltage that is separately activated according to an algorithm to deal with the partial shading condition. Moreover, the resulting optimal control rule involved partial derivatives of the output voltage of the solar PV array with respect to the inductor current. Hence, an exact implementation of the proposed scheme depends on the solar PV model details. This motivated us to obtain a model-free-based control scheme. Moreover, we proposed a piecewise learning-based control that relaxes the need for the exact dynamics. The simulation results, obtained by the implementation of the proposed algorithm on a realistic solar PV model, demonstrated the applicability of the approach in uniform and nonuniform insolation. Furthermore, in the comparison results, the NOC illustrated a suitable convergence response with the minimum oscillations around the MPP. Compared to the second-order SMC, the chattering phenomenon that emerges in the SMC was further decreased by employing the optimal approach with a guaranteed performance measure.

Bibliography

Dennis S. Bernstein. Nonquadratic cost and nonlinear feedback control. *International Journal of Robust and Nonlinear Control*, 3(3):211–229, 1993.

Enrico Bianconi, Javier Calvente, Roberto Giral, Emilio Mamarelis, Giovanni Petrone, Carlos Andrés Ramos-Paja, Giovanni Spagnuolo, and Massimo Vitelli. A fast current-based MPPT technique employing sliding mode control. *IEEE Transactions on Industrial Electronics*, 60(3):1168–1178, 2013.

Ali Bidram, Ali Davoudi, and Robert S. Balog. Control and circuit techniques to mitigate partial shading effects in photovoltaic arrays. *IEEE Journal of Photovoltaics*, 2(4):532–546, 2012.

Canadian Solar. Datasheet for Dymond CS6X-M-FG module, 2022. URL https://si-datastore.s3.us-west-2.amazonaws.com/documents/ooSrMSYUtK69SKhxerIcly4u3wNETGXc6IupV1Tv.pdf.

Giuseppe Carannante, Ciro Fraddanno, Mario Pagano, and Luigi Piegari. Experimental performance of MPPT algorithm for photovoltaic sources subject to inhomogeneous insolation. *IEEE Transactions on Industrial Electronics,* 56(11):4374–4380, 2009.

Chian-Song Chiu and Ya-Lun Ouyang. Robust maximum power tracking control of uncertain photovoltaic systems: A unified TS fuzzy model-based approach. *IEEE Transactions on Control Systems Technology,* 19(6):1516–1526, 2011.

Chen-Chi Chu and Chieh-Li Chen. Robust maximum power point tracking method for photovoltaic cells: A sliding mode control approach. *Solar Energy,* 83(8):1370–1378, 2009.

G. Silva Deaecto, José Cláudio Geromel, F. S. Garcia, and J. A. Pomilio. Switched affine systems control design with application to DC-DC converters. *IET Control Theory and Applications,* 4(7):1201–1210, 2010.

Trishan Esram and Patrick L. Chapman. Comparison of photovoltaic array maximum power point tracking techniques. *IEEE Transactions on Energy Conversion,* 22(2):439–449, 2007.

Milad Farsi and Jun Liu. Nonlinear optimal feedback control and stability analysis of solar photovoltaic systems. *IEEE Transactions on Control Systems Technology,* 28(6):2104–2119, 2019.

Nicola Femia, Giovanni Petrone, Giovanni Spagnuolo, and Massimo Vitelli. Optimization of perturb and observe maximum power point tracking method. *IEEE Transactions on Power Electronics,* 20(4):963–973, 2005.

Wassim M. Haddad and VijaySekhar Chellaboina. *Nonlinear Dynamical Systems and Control: A Lyapunov-Based Approach.* Princeton University Press, 2011.

Wassim M. Haddad and Andrea L'Afflitto. Finite-time stabilization and optimal feedback control. *IEEE Transactions on Automatic Control,* 61(4):1069–1074, 2016.

Young-Hyok Ji, Doo-Yong Jung, Jun-Gu Kim, Jae-Hyung Kim, Tae-Won Lee, and Chung-Yuen Won. A real maximum power point tracking method for mismatching compensation in PV array under partially shaded conditions. *IEEE Transactions on Power Electronics,* 26(4):1001–1009, 2011.

A. Kchaou, A. Naamane, Y. Koubaa, and N. M'sirdi. Second order sliding mode-based MPPT control for photovoltaic applications. *Solar Energy,* 155:758–769, 2017.

Kenji Kobayashi, Ichiro Takano, and Yoshio Sawada. A study on a two stage maximum power point tracking control of a photovoltaic system under partially shaded insolation conditions. In *Proceedings of the IEEE Power Engineering Society General Meeting,* volume 4, pages 2612–2617. IEEE, 2003.

Eftichios Koutroulis and Frede Blaabjerg. A new technique for tracking the global maximum power point of pv arrays operating under partial-shading conditions. *IEEE Journal of Photovoltaics,* 2(2):184–190, 2012.

Jae Ho Lee, HyunSu Bae, and Bo Hyung Cho. Advanced incremental conductance MPPT algorithm with a variable step size. In *Proceedings of the International Power Electronics and Motion Control Conference,* pages 603–607. IEEE, 2006.

Xiao Li, Yaoyu Li, John E Seem, and Peng Lei. Detection of internal resistance change for photovoltaic arrays using extremum-seeking control MPPT signals. *IEEE Transactions on Control Systems Technology*, 24(1):325–333, 2016.

Abderraouf Messai, Adel Mellit, A. Guessoum, and S. A. Kalogirou. Maximum power point tracking using a GA optimized fuzzy logic controller and its FPGA implementation. *Solar Energy*, 85(2):265–277, 2011.

Alivarani Mohapatra, Byamakesh Nayak, Priti Das, and Kanungo Barada Mohanty. A review on MPPT techniques of PV system under partial shading condition. *Renewable and Sustainable Energy Reviews*, 80:854–867, 2017.

P. Moylan and B. Anderson. Nonlinear regulator theory and an inverse optimal control problem. *IEEE Transactions on Automatic Control*, 18(5):460–465, 1973.

Abdollah Noori, Milad Farsi, and Reza Mahboobi Esfanjani. Robust switching strategy for buck-boost converter. In *Proceedings of the International Conference on Computer and Knowledge Engineering*, pages 492–496. IEEE, 2014.

Abdollah Noori, Milad Farsi, and Reza Mahboobi Esfanjani. Design and implementation of a robust switching strategy for DC-DC converters. *IET Power Electronics*, 9(2):316–322, 2016.

Hiren Patel and Vivek Agarwal. MATLAB-based modeling to study the effects of partial shading on PV array characteristics. *IEEE Transactions on Energy Conversion*, 23(1):302–310, 2008.

Raseswari Pradhan and Bidyadhar Subudhi. Double integral sliding mode MPPT control of a photovoltaic system. *IEEE Transactions on Control Systems Technology*, 24(1):285–292, 2016.

J. Prasanth Ram, T. Sudhakar Babu, and N. Rajasekar. A comprehensive review on solar PV maximum power point tracking techniques. *Renewable and Sustainable Energy Reviews*, 67:826–847, 2017.

Miloud Rezkallah, Shailendra Kumar Sharma, Ambrish Chandra, Bhim Singh, and Daniel R. Rousse. Lyapunov function and sliding mode control approach for the solar-PV grid interface system. *IEEE Transactions on Industrial Electronics*, 64(1):785–795, 2017.

Hamza Sahraoui, Larbi Chrifi, Said Drid, and Pascal Bussy. Second order sliding mode control of DC-DC converter used in the photovoltaic system according an adaptive MPPT. *International Journal of Renewable Energy Research*, 6(2), 2016.

P. Sivakumar, Abdullah Abdul Kader, Yogeshraj Kaliavaradhan, and M. Arutchelvi. Analysis and enhancement of PV efficiency with incremental conductance MPPT technique under non-linear loading conditions. *Renewable Energy*, 81:543–550, 2015.

Filippo Spertino and Jean Sumaili Akilimali. Are manufacturing $I–V$ mismatch and reverse currents key factors in large photovoltaic arrays? *IEEE Transactions on Industrial Electronics*, 56(11):4520–4531, 2009.

Kinattingal Sundareswaran, Sankar Peddapati, and Sankaran Palani. Application of random search method for maximum power point tracking in partially shaded photovoltaic systems. *IET Renewable Power Generation*, 8(6):670–678, 2014.

Siew-Chong Tan, Y. M. Lai, and K. Tse Chi. Indirect sliding mode control of power converters via double integral sliding surface. *IEEE Transactions on Power Electronics*, 23(2):600–611, 2008.

Boutabba Tarek, Drid Said, and M. E. H. Benbouzid. Maximum power point tracking control for photovoltaic system using adaptive neuro-fuzzy "ANFIS". In *Proceedings of the International Conference and Exhibition on Ecological Vehicles and Renewable Energies*, pages 1–7, 2013.

Huan-Liang Tsai. Insolation-oriented model of photovoltaic module using Matlab/Simulink. *Solar Energy*, 84(7):1318–1326, 2010.

Weidong Xiao and William G. Dunford. A modified adaptive hill climbing MPPT method for photovoltaic power systems. In *Proceedings of the IEEE Power Electronics Specialists Conference*, pages 1957–1963. IEEE, 2004.

9

An Application to Low-level Control of Quadrotors

This chapter presents a case study on the Structured Online Learning (SOL)-based control of quadrotors. The results presented in this chapter are published in Farsi and Liu [2022].

9.1 Introduction

Quadrotors have received enormous attention because of their efficacy in various applications. Nowadays, quadrotors are produced at a reasonable cost and in different sizes that justify their increasing deployment in new environments. Their applications can range from industry to everyday life and they can reach and operate in situations that may be expensive or dangerous for humans to enter. In response to the high demand, researchers have developed extensive approaches over the past decade to control quadrotors for complicated tasks and high-precision acrobats.

For an effective fly of the quadrotor, there are two levels of controls involved: the low-level control that is required for the stability of the hovering position, and the high-level control that provides a sequence of setpoints as commands to achieve a particular objective. Even though effective controllers have been already implemented on quadrotors for various objectives, the design and tuning procedure of such controllers still requires a considerable amount of expert knowledge about the governing dynamics and experiments to determine the exact system parameters. This becomes even more challenging considering the underactuated and rapid dynamics of the quadrotor. Hence, having successful learning approaches are extremely desired, because it can automate and accelerate the procedure of reaching the flying state with minimum knowledge of the dynamics.

Model-Based Reinforcement Learning: From Data to Continuous Actions with a Python-based Toolbox,
First Edition. Milad Farsi and Jun Liu.

In Dulac-Arnold et al. [2019], the main challenges in the application of Reinforcement Learning (RL) approaches are highlighted. Among RL approaches, Model-based Reinforcement Learning (MBRL) techniques are sometimes preferred over model-free methods, because of their effectiveness in learning from limited data and their computationally tractable properties, which allow real-time inference of the policy. In contrast, direct RL approaches, on the other hand, usually require a large amount of data and a long time of training [Duan et al., 2016]. Therefore, in this chapter, we will only discuss MBRL strategies that provide an opportunity for learning in a few tries.

There are plenty of MBRL approaches implemented on aerial vehicles that use demonstrations led by humans to collect data and realize a model that can best predict the future state for a given action [Coates et al., 2009; Bansal et al., 2016]. Moreover, MBRL techniques are effectively implemented on quadrotors that assume an on-board stable low-level controller to learn the high-level control [Liu et al., 2016a; Abdolhosseini et al., 2013; Becker-Ehmck et al., 2020], which is not in the scope of this chapter. Hence, in this chapter, we are interested in obtaining a low-level control that requires no initial controller and no knowledge of the system parameters.

Recently, the authors of Lambert et al. [2019] suggested an interesting approach in learning a low-level controller by running an MBRL approach on a real nano-quadrotor, best known as Crazyflie. In this approach, they train a neural network by collecting experimental data, which is then used to run a random-shooter Model Predictive Control (MPC) on a graphic processing unit (GPU) to establish a real-time controller. The approach can successfully reach a hovering position in a few tries. However, the need for GPU to learn a low-level control can be seen as a limiting factor of its implementation.

In this chapter, similar to Lambert et al. [2019], we use the flight data obtained along random open-loop trajectories to establish an initial model, with no need for any expert demonstrations. However, in a different approach than [Lambert et al., 2019], once an initial model together with the corresponding controller is obtained, we switch to learning in a closed-loop form to refine the model and the performance achieved. To verify the method, we acquire data from the nonlinear model of the quadrotor, treated as a black box.

For a practical framework, the MBRL approach has to be data-efficient while being fast enough to allow real-time implementation. Unlike [Lambert et al., 2019], our approach does not demand a lot of computational efforts considering that we learn the system in terms of a limited number of bases and accordingly obtain a feedback control rule. Hence, it can be used as a lightweight alternative in implementations. Moreover, in Lambert et al. [2019], the objective is to reach the hovering position, whereas in this chapter, in addition to the attitude control, we also control the position. This means the quadrotor simultaneously learns to reach and stay at a given point in the 3D space. This will also minimize the instances where the quadrotor slides out of the training environment.

For learning purpose, we implemented the SOL approach proposed in Farsi and Liu [2020] with a Recursive Least Squares (RLS) algorithm that is well known for its high efficiency in online applications. Successful applications of RLS can be found in Liu et al. [2016b], Wu et al. [2015], Yang et al. [2013], and Schreier [2012]. In this chapter, as an alternative to the neural network approach, we used a system structured in terms of a library of bases. Accordingly, by sampling the input and state, we employ RLS to update the system model. Then, by exploiting the structure assumed in the model and a quadratic parametrization of the value function in terms of the same set of bases, we obtain a matrix differential equation to update the controller that can be efficiently integrated online. An extension of SOL is presented in Farsi and Liu [2021] for tracking unknown systems that can be also implemented on the quadrotor. However, in this chapter, we will only focus on the attitude-position control.

The rest of the chapter is organized as follows: In Section 9.2, we will introduce the nonlinear model of the quadrotor. In Section 9.3, we will highlight the SOL framework, together with the practical considerations required for a real-time implementation on the quadrotor. Section 9.4 contains the simulation results.

9.2 Quadrotor Model

The nonlinear dynamics of the quadrotor can be written as follows:

$$\dot{y} = v,$$

$$\dot{v} = \begin{bmatrix} 0 \\ 0 \\ -g \end{bmatrix} + \frac{1}{m} R \begin{bmatrix} 0 \\ 0 \\ T \end{bmatrix},$$

$$\dot{R} = R Q(\omega),$$

$$\dot{\omega} = J^{-1}(-\omega \times J\omega + \tau), \tag{9.1}$$

where the states include the 3D position y, the linear velocity v of center of gravity in the inertial frame, the rotation matrix R, and the angular velocity ω in the body frame with respect to the inertial frame. It should be noted that the third equation is written in a matrix form, where R takes value in the special orthogonal group $SO(3) = \{R \in \mathbb{R}^{3 \times 3} | R^{-1} = R^T, \det(R) = 1\}$. Accordingly, the attitude of the quadrotor $\chi = \begin{bmatrix} \phi & \theta & \psi \end{bmatrix}$ can be extracted from R at any time instance, which contains the roll, pitch, and yaw angles, respectively.

Moreover, the gravity acceleration, the body mass, and the inertia matrix are given by g, m, and

$$J = \begin{bmatrix} I_{xx} & -I_{xy} & -I_{xz} \\ -I_{xy} & I_{yy} & -I_{yz} \\ -I_{xz} & -I_{yz} & I_{zz} \end{bmatrix}.$$

The skew-matrix Q is composed by the angular velocity $\omega = \begin{bmatrix} p & q & r \end{bmatrix}^T$ as

$$Q(\omega) = \begin{bmatrix} 0 & -r & q \\ r & 0 & -p \\ -q & p & 0 \end{bmatrix}.$$

The inputs of this system are given by the moments in the body frame

$$\tau = \begin{bmatrix} C_T d(-\bar{\omega}_2^2 - \bar{\omega}_4^2 + \bar{\omega}_1^2 + \bar{\omega}_3^2) \\ C_T d(-\bar{\omega}_1^2 + \bar{\omega}_2^2 + \bar{\omega}_3^2 - \bar{\omega}_4^2) \\ C_D(\bar{\omega}_2^2 + \bar{\omega}_4^2 - \bar{\omega}_1^2 - \bar{\omega}_3^2) \end{bmatrix},$$

and the thrust

$$T = C_T(\bar{\omega}_1^2 + \bar{\omega}_2^2 + \bar{\omega}_3^2 + \bar{\omega}_4^2)$$

generated in the body frame by the rotors, where $\bar{\omega}_i$, d, C_T, and C_D denote the rotational speed of each rotor, the arm length, the lift, and the drag coefficients of the propellers, respectively.

The system is usually actuated by four DC motors that are controlled by Pulse-Width Modulation (PWM) signals. To convert the PWM values to the rotational speed RPM, we use

$$\bar{\omega}_i = \eta_1 u_i + \eta_2,$$

where η_1 and η_2 are the coefficients specified for any motor, and $i \in \{1, 2, 3, 4\}$. Accordingly, in the learning process, we will consider u_i as the inputs of the quadrotor system.

9.3 Structured Online Learning with RLS Identifier on Quadrotor

In what follows, we discuss different stages of the proposed learning procedure in detail. Moreover, we present considerations to be taken into account in the implementation together with the computational properties of the proposed framework.

9.3.1 Learning Procedure

The learning procedure is done in the following order: In the first stage (prerun), we run the quadrotor in an open-loop form with almost equal PWM values for every motor. These values are perturbed slightly in a random way that provides a probing input signal to collect diverse samples from the system. The design and importance of such probing signals are well studied in MBRL techniques, and in system identification algorithms in particular. See, for instance, Pierre et al. [2009] and Kamalapurkar et al. [2018]. The data are used to establish an initial model to start online learning.

In the second stage, we implement the learning in a closed-loop form by using the initial model obtained in the prerun. In the control loop, at any time step t_k, the samples of the states are acquired and a set of bases are evaluated accordingly. Next, by using RLS algorithm, the system model is updated. Then, the measurements and the most recent model coefficients are used to update a value function, which is required to calculate the control value for the next step t_{k+1}.

Remark 9.1 Considering that we assume no knowledge about the system coefficients, initial learning is done through several unsafe tries similar to Lambert et al. [2019]. Hence, the first runs of the system may show poor control performance or instability, which requires a safe training environment and/or a resetting mechanism that allows safe crashes.

Two blocks are responsible for the control and model update in a loop that will be discussed next.

Control Update

Consider the quadrotor model (9.1), where we compose the state vector by using the position, the velocity, the attitude, and the angular velocity as

$$x := \begin{bmatrix} y & v & \chi & \omega \end{bmatrix},$$

where $x \in D \subseteq \mathbb{R}^n$, and $u \in U \subseteq \mathbb{R}^m$ with $n = 12$ and $m = 4$. As mentioned, angles χ can be obtained using the state R.

The cost functional to be minimized along the trajectory started from the initial condition $x_0 = x(0)$ is considered in the following linear quadratic form:

$$J(x_0, u) = \lim_{T \to \infty} \int_0^T e^{-\gamma t} \left(x^T Q_0 x + u^T R_0 u \right) dt, \qquad (9.2)$$

where $Q_0 \in \mathbb{R}^{n \times n}$ is positive semidefinite, $\gamma \geq 0$ is the discount factor, and $R_0 \in \mathbb{R}^{m \times m}$ is a diagonal matrix with only positive values, given by design criteria.

For the closed-loop system, by assuming a feedback control law $u = v(x(t))$ for $t \in [0, \infty)$, the optimal control is given by

$$v^* = \arg \min_{u(\cdot) \in \Gamma(x_0)} J(x_0, u(\cdot)), \tag{9.3}$$

where Γ is the set of admissible controls.

For now, we assume that we can approximate the dynamics of (9.1) in terms of some differentiable bases such as polynomial and trigonometric functions, where the system identification approach will be discussed later. Accordingly, (9.1) is rewritten as follows:

$$\dot{x} = W\Phi(x) + \sum_{j=1}^{m} W_j \Phi(x) u_j, \tag{9.4}$$

where W and $W_j \in \mathbb{R}^{n \times p}$ are the matrices of the coefficients obtained for $j = 1, 2, \ldots, m$, and $\Phi(x) = [\phi_1(x) \quad \ldots \quad \phi_p(x)]^T$ is the set of chosen bases.

In what follows, without loss of generality, the cost defined in (9.2) is transformed to the space of bases $\Phi(x)$, that is

$$J(x_0, u) = \lim_{T \to \infty} \int_0^T e^{-\gamma t} \left(\Phi(x)^T \bar{Q}_0 \Phi(x) + u^T R_0 u \right) dt, \tag{9.5}$$

where $\bar{Q}_0 = \mathrm{diag}\left([Q_0], [\mathbf{0}_{(p-n) \times (p-n)}]\right)$ is a block diagonal matrix that contains all zeros except the first block Q_0, which corresponds to the linear basis x.

Then the corresponding HJB equation can be written by the Hamiltonian defined as follows:

$$-\frac{\partial}{\partial t}(e^{-\gamma t}V)$$

$$= \min_{u(\cdot) \in \Gamma(x_0)} \left\{ H = e^{-\gamma t} \left(\Phi(x)^T \bar{Q}_0 \Phi(x) + u^T R_0 u \right) \right.$$

$$\left. + e^{-\gamma t} \frac{\partial V}{\partial x} \left(W\Phi(x) + \sum_{j=1}^{m} W_j \Phi(x) u_j \right) \right\}. \tag{9.6}$$

In general, there exists no analytical approach that can solve such a partial differential equation and obtain the optimal value function. However, it has been shown in the literature that approximate solutions can be computed by numerical techniques.

Assume the optimal value function in the following form:

$$V = \Phi(x)^T P \Phi(x), \tag{9.7}$$

where P is symmetric. Then the Hamiltonian is given by

$$H = e^{-\gamma t}(\Phi(x)^T \bar{Q}_0 \Phi(x) + u^T R_0 u)$$

$$+ e^{-\gamma t} \Phi(x)^T P \frac{\partial \Phi(x)}{\partial x} \left(W\Phi(x) + \sum_{j=1}^{m} W_j \Phi(x) u_j \right)$$

$$+ e^{-\gamma t}\left(\Phi(x)^T W^T + \sum_{j=1}^{m} u_j^T \Phi(x)^T W_j^T\right) \frac{\partial \Phi(x)}{\partial x}^T P\Phi(x).$$

Moreover, based on the diagonal structure of R_0, the quadratic term of u is rewritten in terms of its components, where $r_{0j} \neq 0$ is the jth component on the diagonal of matrix R_0. To minimize the resulting Hamiltonian we need

$$\frac{\partial H}{\partial u_j} = 2r_{0j}u_j + 2\Phi(x)^T P\frac{\partial \Phi(x)}{\partial x}W_j\Phi(x) \tag{9.8}$$

$$= 0, \qquad j = 1, 2, \dots, m.$$

Hence, the jth optimal control input is obtained as follows:

$$u_j^* = -\Phi(x)^T r_{0j}^{-1} P\frac{\partial \Phi(x)}{\partial x}W_j\Phi(x). \tag{9.9}$$

Based on Farsi and Liu [2020], the following update rule can be obtained by plugging in the optimal control in the Hamiltonian.

$$-\dot{P} = \bar{Q}_0 + P\frac{\partial \Phi(x)}{\partial x}W + W^T\frac{\partial \Phi(x)}{\partial x}^T P - \gamma P$$

$$- P\frac{\partial \Phi(x)}{\partial x}\left(\sum_{j=1}^{m} W_j\Phi(x)r_{0j}^{-1}\Phi(x)^T W_j^T\right)\frac{\partial \Phi(x)}{\partial x}^T P. \tag{9.10}$$

In a standard optimal control approach, this equation has to be solved backward in time, which assumes complete knowledge of the system, i.e. W and W_j along the time horizon. However, in this chapter, we are interested in the learning problem, where the system model may not be known initially. Therefore, we propagate the obtained differential equation in the forward direction. This will provide an opportunity to update our estimation of the system dynamics online at any step together with the control update.

In this approach, we run the system from some $x_0 \in D$, then solve the matrix differential equation (9.10) along the trajectories of the system. Different solvers are already developed that can efficiently integrate differential equations. Although the solver may take smaller steps, we only allow the measurements and control update at time steps $t_k = kh$, where h is the sampling time and $k = 0, 1, 2, \dots$. For solving (9.10) in continuous time, we use the LSODA solver [Hindmarsh and Petzold, 2005], where the weights and the states in this equation are updated by a system identification algorithm and the measurements x_k at each iteration of the control loop, respectively. A recommended choice for P_0 is a matrix with components of zero or very small values.

The differential equation (9.10) also requires evaluations of $\partial \Phi/\partial x_k$ at any time step. Since the bases Φ are chosen beforehand, the partial derivatives can be analytically calculated and stored as functions. Hence, they can be evaluated for any x_k in a similar way as Φ itself. By solving (9.10), we can calculate the control update at any time step t_k according to (9.9). Although at the very first steps of learning, control is not expected to take effective steps toward the control objective, it can help in the exploration of the state space and gradually improve by learning more about the dynamics.

Model Update

In the previous step, we considered a given structured nonlinear system as in (9.4). Therefore, having the control and state samples of the system, we need an algorithm that updates the estimation of system weights. As studied in Brunton et al. [2016] and Kaiser et al. [2018], SINDy is a data-efficient tool to extract the underlying sparse dynamics of the sampled data. In this approach, along with the identification, the sparsity is also promoted in the weights by minimizing

$$[\hat{W} \ \hat{W}_1 \ ... \ \hat{W}_m]_k = \arg\min_{\bar{W}} \ \|\dot{X}_k - \bar{W}\Theta_k\|_2^2 + \lambda \ \|\bar{W}\|_1, \tag{9.11}$$

where k is the time step, $\lambda > 0$, and Θ_k includes a matrix of samples with the columns of

$$\Theta_k^s = [\Phi^T(x^s) \quad \Phi^T(x^s)u_1^s \quad ... \quad \Phi^T(x^s)u_m^s]_k^T,$$

for sth sample. In the same order, $\dot{X}$ keeps a table of sampled state derivatives.

Updating $\hat{W}_k$ based on a history of samples may not be favored as the number of samples needed tends to be large. Especially, real-time implementations may not be possible because of the latency caused by the computations. There exist other techniques that can be alternatively used in different situations, such as neural networks, nonlinear regression, or any other function approximation, and system identification methods. For real-time control applications, considering the linear dependence on the system weights in (9.4), one may choose RLS algorithm that only uses the latest sample of the system and $\hat{W}_{k-1}$, hence will run considerably faster.

For this reason, in the application of SOL on the quadrotor, we employ the RLS algorithm. Moreover, to improve the runtime of the identification we only choose the linear and the constant bases. Considering that the quadrotor usually operates around the hovering situation such an approximation will still be able to preserve the required properties of the system. However, for better results, one can add higher-order polynomial in the library as shown in Farsi and Liu [2020].

In the implementations of SOL done in this chapter, we compare the prediction error $\dot{e}_k = \|\dot{x}_k - \hat{\dot{x}}_k\|$ with the average $\bar{e}_k = \sum_{i=1}^{k} \dot{e}_k / k$. Hence, if the condition $\dot{e}_k > \eta \bar{e}_k$ holds, we use that sample to update the model, where the constant $0 < \eta < 1$ adjusts the threshold. Choosing smaller values of η will increase the rate of adding samples to the database.

9.3.2 Asymptotic Convergence with Uncertain Dynamics

Having arranged the identifier and the controller, it only remains for us to consider the model uncertainty's effect on the asymptotic convergence to the equilibrium point. Assume the system structured as follows:

$$\dot{x} = W\Phi + \sum_{j=1}^{m} W_j \Phi \hat{u}_j + \epsilon, \tag{9.12}$$

where $\hat{u}_j = -\Phi^T R^{-1} \hat{P} \Phi_x \hat{W}_j \Phi$ is the feedback control rule obtained based on the estimation of the system $(\hat{W}, \hat{W}_j)$. Moreover, ϵ is the bounded approximation error in D. By assuming $W = \hat{W} + \tilde{W}$ and $W_j = \hat{W}_j + \tilde{W}_j$, this can be rewritten as follows:

$$\dot{x} = \hat{W}\Phi + \sum_{j=1}^{m} \hat{W}_j \Phi \hat{u}_j + \Delta(t), \tag{9.13}$$

where unidentified dynamics are lumped together as $\Delta(t)$. By the assumption that the feedback control u_j is bounded in D, we have $\|\Delta(t)\| \leq \bar{\Delta}$. For asymptotic convergence, and also to promote the robustness of the controller, the effect of the uncertainty should be taken into account. Hence, we use an auxiliary vector ρ to get

$$\dot{x} = \hat{W}\Phi + \sum_{j=1}^{m} \hat{W}_j \Phi \hat{u}_j + \Delta(t) + \rho - \rho$$

$$= \hat{W}_\rho \Phi + \sum_{j=1}^{m} \hat{W}_j \Phi \hat{u}_j + \Delta(t) - \rho,$$

where assuming that Φ also includes the constant basis, we adjusted the corresponding column in the system matrix to get $\hat{W}_\rho$. In the case $\bar{\Delta} = 0$, the controller $\hat{u}$ can be obtained such that the closed system is locally asymptotically stable. For the case $\bar{\Delta} > 0$, although the system will stay stable for small enough $\bar{\Delta}$, it may not asymptotically converge to zero. Then, similar to Xian et al. [2004] and Qu and Xu [2002], we obtain ρ as below to help to slide the system state to zero

$$\rho = \int_0^t [k_1 x(\tau) + k_2 \operatorname{sign}(x(\tau))] d\tau,$$

where k_1 and k_2 are positive scalars. It can be shown that over time $\|\Delta(t) - \rho\| \to 0$, and hence the system will asymptotically converge to the origin.

9.3.3 Computational Properties

The computational complexity of updating parameters by relation (9.10) is bounded by the complexity of matrix multiplications of dimension p which is

$\mathcal{O}(p^3)$. Moreover, it should be noted that, regarding the symmetry in the matrix of parameters P, this equation updates $L = (p^2 + p)/2$ number of parameters which correspond to the number of bases used in the value function. Therefore, in terms of the number of parameters, the complexity is $\mathcal{O}(L^{3/2})$. It is discussed in Farsi and Liu [2020] that this can be done considerably faster than similar MBRL techniques, such as Kamalapurkar et al. [2016b], Bhasin et al. [2013], and Kamalapurkar et al. [2016a]. If only linear and constant bases are chosen, we will require matrix multiplications of dimension 13 to update the controller for the quadrotor. The runtime results reported in Figure 9.1 for the quadrotor indicates a maximum process time of 8 ms for calculating the control.

Moreover, as mentioned RLS is already implemented in many online identification techniques [Liu et al., 2016b; Wu et al., 2015; Yang et al., 2013; Schreier, 2012]. Similarly, the runtime results in Figure 9.1 confirm that RLS updates can be efficiently done under 2 ms.

Accordingly, the total latency added by the computations in the control loop will be at most about 10 ms. Therefore, the control frequency of 100 Hz is achievable, which is enough for controlling a wide range of quadrotors.

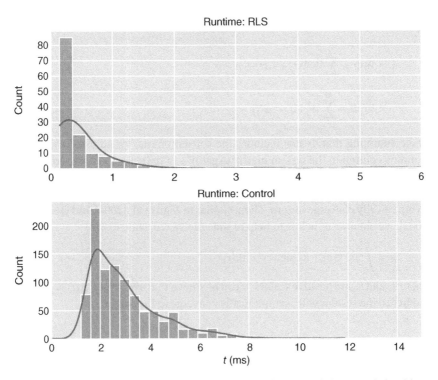

Figure 9.1 The histogram of the runtime of the identification and the control algorithms.

Table 9.1 The coefficients of the simulated Crazyflie.

m	0.33 (Kg)	d	39.73×10^{-3} (m)
I_{xx}	1.395×10^{-5} (Kg×m²)	C_T	0.2025
I_{yy}	1.436×10^{-5} (Kg×m²)	C_D	0.11
I_{zz}	2.173×10^{-5} (Kg×m²)	g	0.98 (m/s²)

9.4 Numerical Results

In the simulation, we consider the nonlinear model (9.1) for the Crazyflie, where the parameters are taken from Luis and Ny [2016] as listed in Table 9.1. The model is treated as a black box to simulate the real-world implementation. This model is integrated by using a Runge–Kutta solver.

Through these simulations, we assume full access to the states. We then obtained the state derivatives using a one-step backward approximation.

We performed the simulations in Python on a 2.6 GHz Intel Core i5, including the 3D graphics generated via the Vpython module [Scherer et al., 2000], as shown in Figure 9.2. The sampling rate is 66.6 Hz ($h = 15$ ms) for all the simulations. Accordingly, the control input value is calculated with the same rate. In the learning process, to simulate the exact behavior of Crazyflie, we integrate the continuous differential equations in sufficiently high precision. However, we only allow measurements at time steps conforming to the sampling rate. Moreover, the initial position is randomly chosen while the rest of states are set to zero.

Figure 9.2 A video of the training procedure can be found at https://youtu.be/QO8Ql83qKFM, where the objective is to learn to fly and reach to the reference position and yaw.

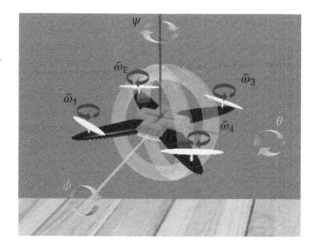

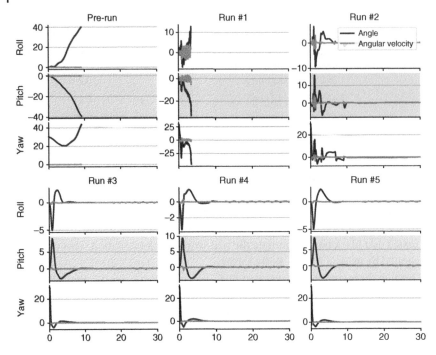

Figure 9.3 The attitude control results in the learning procedure illustrated in different runs.

The controller algorithm is run with the following settings defining the objective functions and the bases

$$Q_0 = \text{diag}([20, 20, 20, 0.8, 0.8, 0.8, 0, 0, 0, 0.4, 0.4, 0.04]),$$

$$R_0 = 3.5 \times 10^{-7} I_{4 \times 4}, \Phi = \{1, x\}, \gamma = 0.4.$$

We choose $x_{\text{ref}} = [0, 0, 3]$ m and $\psi_{\text{ref}} = 30°$. In Figures 9.3 and 9.4, the attitude and position error of the quadrotor with respect to the reference are illustrated within different runs, where we employed three preruns before starting the closed-loop learning. A simulation video of the training is also uploaded at https://youtu.be/QO8Ql83qKFM. Although the quadrotor shows unstable behavior in its very first runs as expected according to Remark 9.1, it can achieve and preserve a stable hovering mode soon after. Figure 9.5 denotes the model coefficients identified using RLS. In addition, it can successfully reach the goal point after collecting 634 samples within 68 seconds of flying until the end of Run #2 (Figures 9.6 and 9.7).

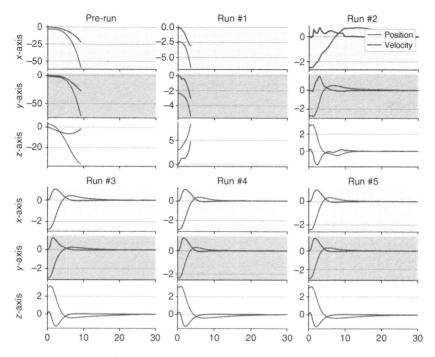

Figure 9.4 The position control results in the learning procedure illustrated in different runs, starting from random initial positions.

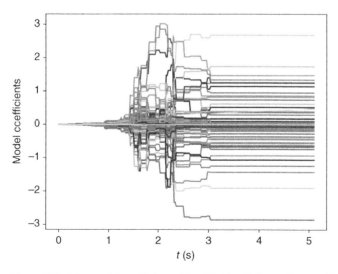

Figure 9.5 The model coefficients, identified by RLS, are shown within a run of the system.

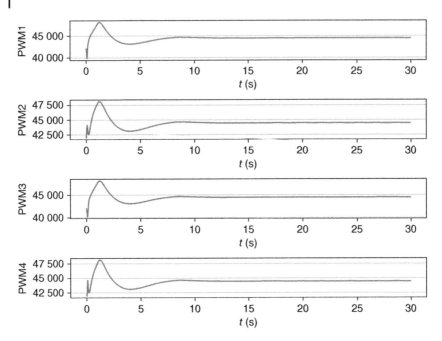

Figure 9.6 The PWM inputs of the quadrotor generated by the learned control.

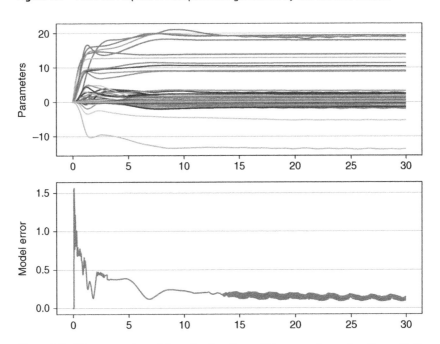

Figure 9.7 The parameters of the value function within a sample run together with the prediction error of the learned model.

9.5 Summary

In this chapter, by focusing on Crazyflie, we implemented the SOL learning algorithm to learn the low-level control for quadrotors to fly and keep the hovering state in a goal position point. To improve the runtime results of the learning, we implemented SOL with the RLS algorithm that is more suitable for online applications. The simulation results, illustrated rapid and efficient learning, where an initial model is obtained by random preruns and then the model was improved within different runs in a closed-loop form. Based on the flight data and runtime results, the approach can be employed to automate the control of the quadrotor. In future work, we will employ the obtained results and the tracking extension of SOL combined with a higher-level approach to achieve more complex objectives with no knowledge of the dynamics.

Bibliography

Mahyar Abdolhosseini, Youmin Zhang, and Camille Alain Rabbath. An efficient model predictive control scheme for an unmanned quadrotor helicopter. *Journal of Intelligent and Robotic Systems*, 70(1–4):27–38, 2013.

Somil Bansal, Anayo K. Akametalu, Frank J. Jiang, Forrest Laine, and Claire J. Tomlin. Learning quadrotor dynamics using neural network for flight control. In *Proceedings of the IEEE Conference on Decision and Control*, pages 4653–4660. IEEE, 2016.

Philip Becker-Ehmck, Maximilian Karl, Jan Peters, and Patrick van der Smagt. Learning to fly via deep model-based reinforcement learning. *arXiv preprint arXiv:2003.08876*, 2020.

Shubhendu Bhasin, Rushikesh Kamalapurkar, Marcus Johnson, Kyriakos G. Vamvoudakis, Frank L. Lewis, and Warren E. Dixon. A novel actor–critic–identifier architecture for approximate optimal control of uncertain nonlinear systems. *Automatica*, 49(1):82–92, 2013.

Steven L. Brunton, Joshua L. Proctor, and J. Nathan Kutz. Discovering governing equations from data by sparse identification of nonlinear dynamical systems. *Proceedings of the National Academy of Sciences of the United States of America*, 113(15):3932–3937, 2016.

Adam Coates, Pieter Abbeel, and Andrew Y. Ng. Apprenticeship learning for helicopter control. *Communications of the ACM*, 52(7):97–105, 2009.

Yan Duan, Xi Chen, Rein Houthooft, John Schulman, and Pieter Abbeel. Benchmarking deep reinforcement learning for continuous control. In *International Conference on Machine Learning*, pages 1329–1338, 2016.

Gabriel Dulac-Arnold, Daniel Mankowitz, and Todd Hester. Challenges of real-world reinforcement learning. *arXiv preprint arXiv:1904.12901*, 2019.

Milad Farsi and Jun Liu. Structured online learning-based control of continuous-time nonlinear systems. *IFAC-PapersOnLine*, 53(2):8142–8149, 2020.

Milad Farsi and Jun Liu. A structured online learning approach to nonlinear tracking with unknown dynamics. In *Proceedings of the American Control Conference*, pages 2205–2211. IEEE, 2021.

Milad Farsi and Jun Liu. Structured online learning for low-level control of quadrotors. In *Proceedings of the American Control Conference*. IEEE, 2022.

A. C. Hindmarsh and L. R. Petzold. LSODA, ordinary differential equation solver for stiff or non-stiff system, 2005.

Eurika Kaiser, J. Nathan Kutz, and Steven L. Brunton. Sparse identification of nonlinear dynamics for model predictive control in the low-data limit. *Proceedings of the Royal Society A*, 474(2219):20180335, 2018.

Rushikesh Kamalapurkar, Joel A. Rosenfeld, and Warren E. Dixon. Efficient model-based reinforcement learning for approximate online optimal control. *Automatica*, 74:247–258, 2016a.

Rushikesh Kamalapurkar, Patrick Walters, and Warren E. Dixon. Model-based reinforcement learning for approximate optimal regulation. *Automatica*, 64(C):94–104, 2016b.

Rushikesh Kamalapurkar, Patrick Walters, Joel Rosenfeld, and Warren Dixon. *Reinforcement Learning for Optimal Feedback Control*. Springer, 2018.

Nathan O. Lambert, Daniel S. Drew, Joseph Yaconelli, Sergey Levine, Roberto Calandra, and Kristofer S J Pister. Low-level control of a quadrotor with deep model-based reinforcement learning. *IEEE Robotics and Automation Letters*, 4(4):4224–4230, 2019.

Hao Liu, Danjun Li, Jianxiang Xi, and Yisheng Zhong. Robust attitude controller design for miniature quadrotors. *International Journal of Robust and Nonlinear Control*, 26(4):681–696, 2016a.

X. Y. Liu, Stefano Alfi, and Stefano Bruni. An efficient recursive least square-based condition monitoring approach for a rail vehicle suspension system. *Vehicle System Dynamics*, 54(6):814–830, 2016b.

Carlos Luis and Jérôme Le Ny. Design of a trajectory tracking controller for a nanoquadcopter. *arXiv preprint arXiv:1608.05786*, 2016.

John W. Pierre, Ning Zhou, Francis K. Tuffner, John F. Hauer, Daniel J. Trudnowski, and William A Mittelstadt. Probing signal design for power system identification. *IEEE Transactions on Power Systems*, 25(2):835–843, 2009.

Zhihua Qu and Jian-Xin Xu. Model-based learning controls and their comparisons using lyapunov direct method. *Asian Journal of Control*, 4(1):99–110, 2002.

David Scherer, Paul Dubois, and Bruce Sherwood. VPython: 3D interactive scientific graphics for students. *Computing in Science & Engineering*, 2(5):56–62, 2000.

Matthias Schreier. Modeling and adaptive control of a quadrotor. In *Proceedings of the IEEE International Conference on Mechatronics and Automation*, pages 383–390. IEEE, 2012.

Lifu Wu, Xiaojun Qiu, Ian S. Burnett, and Yecai Guo. A recursive least square algorithm for active control of mixed noise. *Journal of Sound and Vibration*, 339:1–10, 2015.

Bin Xian, Darren M. Dawson, Marcio S. de Queiroz, and Jian Chen. A continuous asymptotic tracking control strategy for uncertain nonlinear systems. *IEEE Transactions on Automatic Control*, 49(7):1206–1211, 2004.

Jinpeng Yang, Zhihao Cai, Qing Lin, and Yingxun Wang. Self-tuning pid control design for quadrotor uav based on adaptive pole placement control. In *Proceedings of the Chinese Automation Congress*, pages 233–237. IEEE, 2013.

10

Python Toolbox

To implement the learning framework in Chapter 5, we develop a Python toolbox. In this chapter, we highlight the main features of this toolbox. In this toolbox, we use the pendulum system as an example to demonstrate different features of the framework. However, it can be applied on other dynamics with different quadratic objectives given. The codes of this Python toolbox can be accessed at Farsi and Liu [2022].

10.1 Overview

We present the details of the toolbox in three sections. A general view of the toolbox is illustrated in Figure 10.1 that includes these three parts of the tool developed. According to Figure 10.1, the objective and the process are given on the left, that provide the elements we need to define the problem. In the middle, we illustrate the core of the toolbox including classes to handle learning of the model and control through interactions with the system. On the right, the output classes are shown. As seen, circular blocks are categorized in different colors, any of which represents a Python class. In the next sections, we review the methods available for any of these classes in detail. Moreover, through an example, we present the definitions and initializations required for each block.

10.2 User Inputs

In this section, we introduce the classes required for defining the learning problem that include the process and the control objective.

Model-Based Reinforcement Learning: From Data to Continuous Actions with a Python-based Toolbox, First Edition. Milad Farsi and Jun Liu.

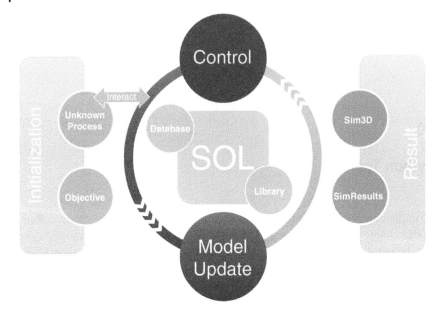

Figure 10.1   A view of the Structured Online Learning (SOL) toolbox is shown. Each colored circle represents a class with a particular function.

10.2.1   Process

Nonlinear systems are defined in the state space form. This system is treated as a black-box process where one can only feed the control input and measure the states at some particular time steps. The dimensions of the model and the input are stored in this class together with the ranges of the domain. Figure 10.2 illustrates the functions involved together with the inputs and outputs of the class.

The followings are the methods available for this class:

- __init__(): We defines the system state, input, and output dimensions. Moreover, the system coefficients are set. The domain of the simulation is chosen by using an array of intervals.
- dynamics(): The dynamics are defined, whereby taking in the control and state it returns an n-dimensional array representing $\dot{x}$.
- integrate(): We set up the integrator. Accordingly, the system is integrated starting from the previous state for an interval of time which is given by the time step h. The solver and its settings can be set by the user to obtain sufficient accuracy and performance.
- read_sensor(): Once the dynamics are integrated, this method generates the appropriate measurements of the states for the process defined earlier. In addition, a noisy measurement can be set up to model realistic measurements.

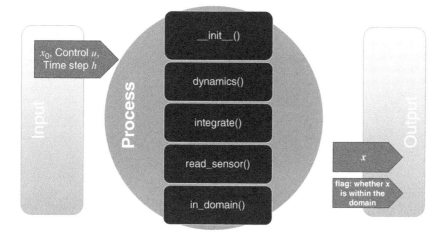

Figure 10.2 The Process class is illustrated including the methods available. The initial state, control, and the time step are given as inputs. Then the state can be measured at the end of each time interval.

- in_domain(): According to the domain defined, this method returns a true value. Otherwise, it reports the dimensions for which the boundaries of the domain are crossed.

10.2.2 Objective

Having defined the nonlinear system as an unknown process of concern, one needs to set objective of learning. According, to the problem formulation, we define the objective using a Linear Quadratic Regulator (LQR) cost. The following methods are used as shown in Figure 10.3

- _init_(): Initialization is done by using Q and R matrices and dimensions of the system.
- stage_cost(). Given the state and control at each time step, we calculate the running cost.

10.3 SOL

In this section, the core part of the toolbox is introduced that includes the model update and control computation procedures.

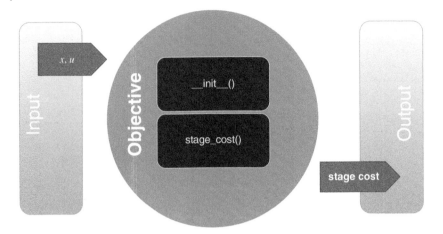

Figure 10.3 The objective class that is defined based on an LQR cost.

10.3.1 Model Update

As shown in Figure 10.4, there exist two classes involved for updating the model. We first discuss the functions of the model update briefly. Then, we introduce the `Database` and `Library` classes.

- `__init__ ()`: We initialize the weights W using the dimension of the system and the list of bases.
- `update()`: We update W using one of the system identification techniques. Depending on the learning algorithm chosen, a history of samples given by `Database` or only the current sample may be employed.
- `evaluate()`: A prediction is made using W and evaluations of bases and gradients. The bases and gradients are given by `Library`, for pair (x,u) that will be explained later.

In what follows, the `Database` and `Library` classes are presented.

10.3.2 Database

In the cases Least Square (LS) or Sparse Identification of Nonlinear Dynamics (SINDy) are selected as the model learner, we need to prepare a batch of input–output data. The dataset includes pairs of sampled approximated state derivatives `X_dot` and set of bases `db_Theta`, in terms of x and u. We use the `Database` class to generate and save the datasets, as shown in Figure 10.5.

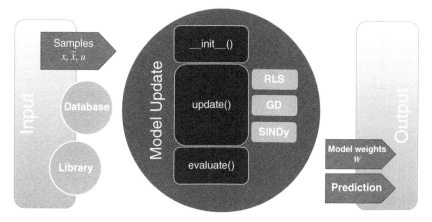

Figure 10.4 A view of the procedure for updating the model is given. The samples in `Database` and `Library` are used as input for updating the model. Then the update model can be used for making predictions.

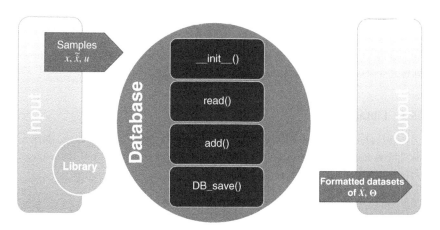

Figure 10.5 The `Database` class are illustrated, where given the samples of the system, the datasets required for updating the model in the batch mode are generated and saved.

This class includes the following methods:

- `__init__()`: We initialize the dataset according to the maximum number of samples set by the user. If there already exists saved data in the directory, they will be loaded instead.
- `read()`: This method slices and returns only the part of datasets that contains the samples stored.

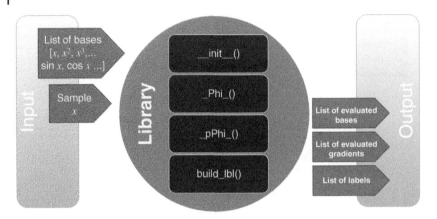

Figure 10.6 Using the samples of the system and the set of bases chosen by the user, the illustrated `Library` class constructs the bases vector, gradient vector, and the corresponding labels for printing purpose.

- `add()`: A column to each of the datasets `X_dot` and `db_Theta` is written that correspond to the current sample given by $x, \tilde{x}, u$. If overflow occurs, the oldest sample is rewritten in the datasets.
- `DB_save()`: Using this method, we save the datasets to the directory in "npy" format so that they can be loaded in next runs if required.

10.3.3 Library

We use this class to handle the operations involving bases and their partial derivatives. These include constructing the set of bases and evaluations of these bases and their gradients as shown in Figure 10.6.

- `__init__()`: Bases and gradient lists are initialized.
- `_Phi_()`: Using this method, we construct and evaluate an array of bases for x according to the dimension of the system and keys chosen for bases.
- `_pPhi_()`: We make and evaluate the gradient matrices corresponding to the list of bases generated by `_Phi_()`.
- `build_lbl()`: A list of labels are built that corresponds to the array of bases generated by `_Phi_()`. This is generated for printing purpose to be used by `SimResults` which will be explained later.

10.3.4 Control

Once the model parameters are updated, we compute the feedback control accordingly. To do this, we feed the updated model parameters to the `Control` class. According to the control algorithm, we need to set up an Ordinary Differential

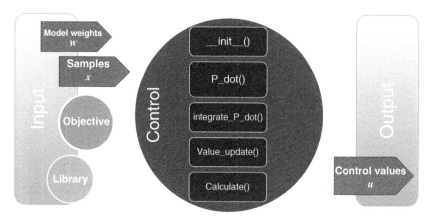

Figure 10.7 The Control class is illustrated, where the Objective and Library, state values and weights of the model are taken as inputs to compute the control value.

Equation (ODE) solver to update the value parameters. In addition, we need evaluations of bases and gradients which can be obtained from Library. A view of Control is shown in Figure 10.7.

- __init__ (). We initialize P and the control value with zeros.
- P_dot(). The matrix differential equation is defined in terms of P. This uses gradients and bases given by the Library.
- integrate_P_dot(). We set up the integrator to update P using P_dot() equation. A solver is chosen and adjusted for this purpose.
- Value_update(). The current value is calculated using updated P and x.
- Calculate(). We calculate the control value using the control rule with the updated P, W, and x.

- __init__ (): Given x_0, Q, and R, we define and initialize the cost.
- stage_cost(). Returns the running cost by using the current state and coefficients defining the objective.

10.4 Display and Outputs

After updating the model and the controller at each iteration, it only remains to keep record of the results for later inspection and visualization at the end of each episode of learning. Accordingly, the analytical model learned can be printed out for inspection. Moreover, different set of graph can be chosen to be generated. In addition, the trajectories may be employed for a graphical simulation. In this section, we provide the tools one can use for generating desirable outputs of the results (Figures 10.8 and 10.9).

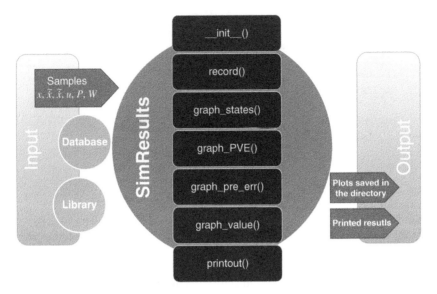

Figure 10.8 The class shown is used to record and visualize the simulation results. The outputs include printed results and graphs of the model and value parameters and prediction error and system responses.

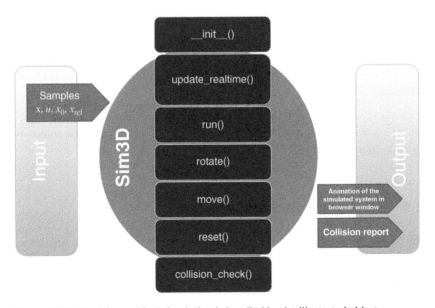

Figure 10.9 The 3D graphical simulation is handled by the illustrated object.

10.4.1 Graphs and Printouts

The methods employed for printing and visualizing the results are reviewed in what follows. For this end, after initialization of the class, we keep records of any variables required to be graphed, and then at the end of each episode, we generate and save all the figures.

- `__init__()`. Using the length of the simulation time, we define arrays of suitable dimensions to keep a record of evolutions of variables for plotting purpose.
- `record()`. We write the current value of variables in the appropriate position of the arrays that correspond to the current step index.
- `graph_states()`. This method draws all of the states and control signals. Two lists of legends and colors are assigned to differentiate them.
- `graph_PVE()`. The evolutions of the value parameters, the value function, and the prediction error are illustrated.
- `graph_pre_err()`: The evolutions of the prediction error are drawn.
- `graph_value()`: We draw the evolutions of the value function.
- `printout()`. The analytical state space model and value function learned are printed out. The model is printed in terms of bases in the state space form. In the same way, the value function is printed in terms of bases. Moreover, a report of each episode including the number of samples is provided.

10.4.2 3D Simulation

In the followings, we explain a typical simulation class is employed to run in 3D using VPython [Scherer et al., 2000]. However, depending on the application, a 3D simulation may or may not be used.

- `__init__()`: We employ `display_instructions()` to print the instruction and information on the output page. Then, the 3D environment is built and initialized using `build_environment()`.
- `update_realtime()`. If online simulation is chosen by a user, this updates the position and rotation of the objects at each time step according to the integrated dynamics. If the simulation is not online taking place at the same time as the learning process, the data are recorded to be played offline at the end of the episode. This helps to resolve lagging in the video caused by computations.
- `run()`: If the simulation is offline, this is used at the end of each episode to play the simulation.
- `rotate()`. Given the difference of the angles between current and the previous steps, this updates the rotation of the 3D object.
- `move()`: Given the difference of the angles between current and the previous steps, this rotates and updates the rotation of the 3D object.

- `reset()`. At the end of each episode, some objects are deleted so that they can be rebuilt in the next episode at the initial state.
- `collision_check()`: Physical constraints in the simulation environment can be defined that trigger the collision report. This can be used to stop the simulation.

10.5 Summary

In this chapter, we presented a toolbox that can be a starting point for developments based on the proposed Model-based Reinforcement Learning (MBRL) technique by providing tools for learning, analysis, and visualization. First, we defined the system and the objective. Second, we discussed how to set up the learning procedure. Third, we reviewed the options available for observing the results through printing and visualization.

Bibliography

Milad Farsi and Jun Liu. Python toolbox for structured online learning-based (SOL) control. https://github.com/Miiilad/SOL-Python-toolbox, 2022. [Online; accessed 4-June-2022].

David Scherer, Paul Dubois, and Bruce Sherwood. VPython: 3D interactive scientific graphics for students. *Computing in Science & Engineering*, 2(5):56–62, 2000.

Appendix

A.1 Supplementary Analysis of Remark 5.4

Consider a ball of radius r around the origin, denoted by D_r. We will show that the dominating part of the solutions of (5.13) is equivalent to (5.19) for some small r. For this purpose, we first consider the Taylor expansion of bases in (5.14) and its partial derivatives:

$$\Phi(x) = \begin{bmatrix} 1 \\ x \\ \Gamma_1 x + O(x^2) \end{bmatrix}, \qquad \frac{\partial \Phi(x)}{\partial x} = \begin{bmatrix} 0 \\ I \\ \Gamma_1 + O(x) \end{bmatrix}, \tag{A.1}$$

where $O(x^\alpha)$ denotes higher-order terms $x_1^{\alpha_1} x_2^{\alpha_2} \dots x_n^{\alpha_n}$ with $\alpha = \sum_{i=1}^{n} \alpha_i$, and non-negative integers α_i. It should be noted that the expansion of the regulated nonlinear bases does not include any constant term according to Remark 5.1. Moreover, $\bar{Q}$ and P are structured matrices including rectangular blocks of appropriate dimensions as follows:

$$\bar{Q} = \begin{bmatrix} 0 & 0 & 0 \\ 0 & Q & 0 \\ 0 & 0 & 0 \end{bmatrix}, \qquad P = \begin{bmatrix} P_1 & P_2 & P_3 \\ P_2^T & P_4 & P_5 \\ P_3^T & P_5^T & P_6 \end{bmatrix}.$$

By substituting $\bar{Q}$, P into (5.13), it is easy to investigate that starting from the initial condition $P(0) = 0$, any term in the equations of $\dot{P}_1$, $\dot{P}_2$, and $\dot{P}_3$ will depend on P_1, P_2, or P_3 as follows:

$$\dot{P}_1 = -P_2 K_1 K_1^T P_2^T - P_3 \Gamma_1 K_1 K_1^T P_2^T - P_2 K_1 K_1^T P_3^T$$
$$- P_3 \Gamma_1 K_1 K_1^T P_3^T,$$

$$\dot{P}_2 = P_2 W_2 + P_3 \Gamma_1 W_2 - P_2 K_1 K_1^T P_4 - P_3 \Gamma_1 K_1 K_1^T P_4$$
$$- P_2 K_1 K_1^T P_5^T - P_3 \Gamma_1 K_1 K_1^T P_5^T,$$

Model-Based Reinforcement Learning: From Data to Continuous Actions with a Python-based Toolbox, First Edition. Milad Farsi and Jun Liu.

$$\dot{P}_3 = P_2 W_3 + P_3 \Gamma_1 W_3 - P_2 K_1 K_1^T P_5 - P_3 \Gamma_1 K_1 K_1^T P_5$$
$$- P_2 K_1 K_1^T P_6 - P_3 \Gamma_1 K_1 K_1^T P_6,$$

where $K_1 = W_{j_1} + W_{j_2} x + W_{j_3} \Gamma_1 x$. Therefore, solutions P_1, P_2, and P_3 will always stay at zeros, and the matrix P will only grow on the block $\begin{bmatrix} P_4 & P_5 \\ P_5^T & P_6 \end{bmatrix}$. Therefore, for brevity, we will follow the computations only for this block as long as this simplification does not cause ambiguity. Now, let us take one step back and start with

$$\Phi(x)^T \dot{P} \Phi(x) = \Phi(x)^T \bar{Q} \Phi(x)$$
$$+ \Phi(x)^T P \frac{\partial \Phi(x)}{\partial x} W \Phi(x) + \Phi(x)^T W^T \frac{\partial \Phi(x)}{\partial x}^T P \Phi(x)$$
$$- \Phi(x)^T P \Phi_x W_j \Phi(x) r_j^{-1} \Phi(x)^T W_j^T \frac{\partial \Phi(x)}{\partial x}^T P \Phi(x)$$
$$- \gamma \Phi(x)^T P \Phi(x), \tag{A.2}$$

as in (5.12). For the nonzero block of P_4, P_5, and P_6 with the corresponding bases, the left-hand side can be rewritten as follows:

$$\begin{bmatrix} 1 \\ x \\ \Gamma_1 x + O(x^2) \end{bmatrix}^T \begin{bmatrix} 0 & 0 & 0 \\ 0 & \dot{P}_4 & \dot{P}_5 \\ 0 & \dot{P}_5^T & \dot{P}_6 \end{bmatrix} \begin{bmatrix} 1 \\ x \\ \Gamma_1 x + O(x^2) \end{bmatrix}$$

$$= \begin{bmatrix} 1 \\ x \\ O(x^2) \end{bmatrix}^T \begin{bmatrix} 0 & 0 & 0 \\ 0 & \dot{P}_4 + \Gamma_1^T \dot{P}_5^T + \dot{P}_5 \Gamma_1 + \Gamma_1^T \dot{P}_6 \Gamma_1 & \dot{P}_5 + \Gamma_1^T \dot{P}_6 \\ 0 & \dot{P}_5^T + \dot{P}_6 \Gamma_1 & \dot{P}_6 \end{bmatrix}$$

$$\begin{bmatrix} 1 \\ x \\ O(x^2) \end{bmatrix}, \tag{A.3}$$

where we shifted the linear term in the third entry of the bases to the second. In the next step, we will consider the following change of variables throughout the matrix differential equation:

$$Z_1 = P_4 + \Gamma_1^T P_5^T + P_5 \Gamma_1 + \Gamma_1^T P_6 \Gamma_1,$$
$$Z_2 = P_5 + \Gamma_1^T P_6,$$
$$Z_3 = P_6. \tag{A.4}$$

For this reason, we apply the same modification of bases to all the terms in the right-hand side of (A.2). The modification will not affect the first term since $\bar{Q}$ is zero everywhere except in the block corresponding to the second basis, which remained unchanged. Then, the second term in the right-hand side of (A.2) becomes

$$\Phi(x)^T P \frac{\partial \Phi(x)}{\partial x} W \Phi(x)$$

$$= \begin{bmatrix} 1 \\ x \\ \Gamma_1 x + O(x^2) \end{bmatrix}^T \begin{bmatrix} 0 & 0 & 0 \\ 0 & P_4 & P_5 \\ 0 & P_5^T & P_6 \end{bmatrix} \begin{bmatrix} 0 \\ \mathbf{I} \\ \Gamma_1 + O(x) \end{bmatrix}$$

$$\begin{bmatrix} 0 & W_2 & W_3 \end{bmatrix} \begin{bmatrix} 1 \\ x \\ \Gamma_1 x + O(x^2) \end{bmatrix}$$

$$= \begin{bmatrix} 1 \\ x \\ \Gamma_1 x + O(x^2) \end{bmatrix}^T \begin{bmatrix} 0 & 0 & 0 \\ 0 & \Upsilon_1 & \Upsilon_2 \\ 0 & \Upsilon_3 & \Upsilon_4 \end{bmatrix} \begin{bmatrix} 1 \\ x \\ \Gamma_1 x + O(x^2) \end{bmatrix}$$

$$= \begin{bmatrix} 1 \\ x \\ O(x^2) \end{bmatrix}^T \begin{bmatrix} 0 & 0 & 0 \\ 0 & \bar{\Upsilon}_1 & \bar{\Upsilon}_2 \\ 0 & \bar{\Upsilon}_3 & \bar{\Upsilon}_4 \end{bmatrix} \begin{bmatrix} 1 \\ x \\ O(x^2) \end{bmatrix}, \tag{A.5}$$

where

$$\Upsilon_1 = P_4 W_2 + P_5 \Gamma_1 W_2 + P_5 O(x) W_2,$$
$$\Upsilon_2 = P_4 W_3 + P_5 \Gamma_1 W_3 + P_5 O(x) W_3,$$
$$\Upsilon_3 = P_5^T W_2 + P_6 \Gamma_1 W_2 + P_6 O(x) W_2,$$
$$\Upsilon_4 = P_5^T W_3 + P_6 \Gamma_1 W_3 + P_6 O(x) W_3,$$
$$\bar{\Upsilon}_1 = Z_1(W_2 + W_3 \Gamma_1) + Z_2 O(x)(W_2 + W_3 \Gamma_1),$$
$$\bar{\Upsilon}_2 = Z_1 W_3 + Z_2 O(x) W_3,$$
$$\bar{\Upsilon}_3 = Z_2^T(W_2 + W_3 \Gamma_1) + Z_3 O(x)(W_2 + W_3 \Gamma_1),$$
$$\bar{\Upsilon}_4 = Z_2^T W_3 + Z_3 O(x) W_3,$$

and we used (A.4) to get

$$P_4 W_2 + P_5 \Gamma_1 W_2 + P_4 W_3 \Gamma_1 + P_5 \Gamma_1 W_3 \Gamma_1$$
$$\quad + \Gamma_1^T P_5^T W_2 + \Gamma_1^T P_6 \Gamma_1 W_2 + \Gamma_1^T P_5^T W_3 \Gamma_1 + \Gamma_1^T P_6 \Gamma_1 W_3 \Gamma_1$$
$$= (P_4 + \Gamma_1^T P_5^T + P_5 \Gamma_1 + \Gamma_1^T P_6 \Gamma_1)(W_2 + W_3 \Gamma_1)$$
$$= Z_1(W_2 + W_3 \Gamma_1),$$
$$P_5 O(x) W_2 + P_6 O(x) W_3 \Gamma_1 + \Gamma_1^T P_6 O(x) W_2 + \Gamma_1^T P_6 O(x) W_3 \Gamma_1$$
$$= Z_2 O(x)(W_2 + W_3 \Gamma_1),$$
$$P_4 W_3 + P_5 \Gamma_1 W_3 + P_5 O(x) W_3 + \Gamma_1^T P_5^T W_3$$
$$\quad + \Gamma_1^T P_6 \Gamma_1 W_3 + \Gamma_1^T P_6 O(x) W_3$$

$$= (P_4 + P_5\Gamma_1 + \Gamma_1^T P_5^T)W_3 + (P_5 + \Gamma_1^T P_6)O(x)W_3$$
$$+ \Gamma_1^T P_6 \Gamma_1 W_3$$
$$= (Z_1 - \Gamma_1^T Z_3 \Gamma 1)W_3 + Z_2 O(x)W_3 + \Gamma_1^T Z_3 \Gamma_1 W_3$$
$$= Z_1 W_3 + Z_2 O(x)W_3,$$

$$P_5^T W_2 + P_6 \Gamma_1 W_2 + P_6 O(x)W_2 + P_5^T W_3 \Gamma_1 + P_6 \Gamma_1 W_3 \Gamma_1$$
$$+ P_6 O(x)W_3 \Gamma_1$$
$$= (P_5^T + P_6 \Gamma_1)(W_2 + W_3 \Gamma_1) + P_6 O(x)(W_2 + W_3 \Gamma_1)$$
$$= Z_2^2(W_2 + W_3 \Gamma_1) + Z_3 O(x)(W_2 + W_3 \Gamma_1),$$

$$P_5^T W_3 + P_6 \Gamma_1 W_3 + P_6 O(x)W_3$$
$$= Z_2^T W_3 + Z_3 O(x)W_3.$$

Furthermore, for the last term in the right-hand side of (A.2), the followings hold:

$$\Phi(x)^T P \frac{\partial \Phi(x)}{\partial x} \sum_{j=1}^{m} (W_j \Phi(x) r_j^{-1} \Phi(x)^T W_j^T) \frac{\partial \Phi(x)}{\partial x}^T P \Phi(x)$$

$$= \begin{bmatrix} 1 \\ x \\ \Gamma_1 x + O(x^2) \end{bmatrix}^T \begin{bmatrix} 0 & 0 & 0 \\ 0 & P_4 & P_5 \\ 0 & P_5^T & P_6 \end{bmatrix} \begin{bmatrix} 0 \\ I \\ \Gamma_1 + O(x) \end{bmatrix}$$

$$\times \sum_{j=1}^{m} \left(\begin{bmatrix} W_{j_1} & W_{j_2} & W_{j_3} \end{bmatrix} \begin{bmatrix} 1 \\ x \\ \Gamma_1 x + O(x^2) \end{bmatrix} r_j^{-1} \begin{bmatrix} 1 \\ x \\ \Gamma_1 x + O(x^2) \end{bmatrix}^T \begin{bmatrix} W_{j_1} \\ W_{j_2} \\ W_{j_3} \end{bmatrix} \right)$$

$$\times \begin{bmatrix} 0 \\ I \\ \Gamma_1 + O(x) \end{bmatrix}^T \begin{bmatrix} 0 & 0 & 0 \\ 0 & P_4 & P_5 \\ 0 & P_5^T & P_6 \end{bmatrix} \begin{bmatrix} 1 \\ x \\ \Gamma_1 x + O(x^2) \end{bmatrix}$$

$$= \begin{bmatrix} 1 \\ x \\ \Gamma_1 x + O(x^2) \end{bmatrix}^T \begin{bmatrix} 0 & 0 & 0 \\ 0 & P_4 & P_5 \\ 0 & P_5^T & P_6 \end{bmatrix} \begin{bmatrix} 0 \\ I \\ \Gamma_1 + O(x) \end{bmatrix} \Omega_2$$

$$\times \begin{bmatrix} 0 \\ I \\ \Gamma_1 + O(x) \end{bmatrix}^T \begin{bmatrix} 0 & 0 & 0 \\ 0 & P_4 & P_5 \\ 0 & P_5^T & P_6 \end{bmatrix} \begin{bmatrix} 1 \\ x \\ \Gamma_1 x + O(x^2) \end{bmatrix}$$

$$= \begin{bmatrix} 1 \\ x \\ \Gamma_1 x + O(x^2) \end{bmatrix}^T \left(\begin{bmatrix} 0 \\ P_4 + P_5\Gamma_1 \\ P_5^T + P_6\Gamma_1 \end{bmatrix} + \begin{bmatrix} 0 \\ P_5 O(x) \\ P_6 O(x) \end{bmatrix} \right) \Omega_2$$

$$\times \left(\begin{bmatrix} 0 \\ P_4 + P_5\Gamma_1 \\ P_5^T + P_6\Gamma_1 \end{bmatrix} + \begin{bmatrix} 0 \\ P_5 O(x) \\ P_6 O(x) \end{bmatrix} \right)^T \begin{bmatrix} 1 \\ x \\ \Gamma_1 x + O(x^2) \end{bmatrix}$$

$$
= \begin{bmatrix} 1 \\ x \\ \Gamma_1 x + O(x^2) \end{bmatrix}^T \begin{bmatrix} 0 & 0 & 0 \\ 0 & \Omega_3 & \Omega_4 \\ 0 & \Omega_4^T & \Omega_5 \end{bmatrix} \begin{bmatrix} 1 \\ x \\ \Gamma_1 x + O(x^2) \end{bmatrix}
$$

$$
+ \begin{bmatrix} 1 \\ x \\ \Gamma_1 x + O(x^2) \end{bmatrix}^T \begin{bmatrix} 0 & 0 & 0 \\ 0 & \Omega_6 & \Omega_7 \\ 0 & \Omega_7^T & \Omega_8 \end{bmatrix} \begin{bmatrix} 1 \\ x \\ \Gamma_1 x + O(x^2) \end{bmatrix}
$$

$$
= \begin{bmatrix} 1 \\ x \\ O(x^2) \end{bmatrix}^T \begin{bmatrix} 0 & 0 & 0 \\ 0 & \bar{\Omega}_3 & \bar{\Omega}_4 \\ 0 & \bar{\Omega}_4^T & \bar{\Omega}_5 \end{bmatrix} \begin{bmatrix} 1 \\ x \\ O(x^2) \end{bmatrix} + \begin{bmatrix} 1 \\ x \\ O(x^2) \end{bmatrix}^T
$$

$$
\times \begin{bmatrix} 0 & 0 & 0 \\ 0 & \bar{\Omega}_6 & \bar{\Omega}_7 \\ 0 & \bar{\Omega}_7^T & \bar{\Omega}_8 \end{bmatrix} \begin{bmatrix} 1 \\ x \\ O(x^2) \end{bmatrix}, \tag{A.6}
$$

where Ω_2 to Ω_8 are defined as follows:

$$
\Omega_2 = \sum_{j=1}^{m} r_j^{-1} \begin{bmatrix} W_{j_1} & W_{j_2} & W_{j_3} \end{bmatrix} \begin{bmatrix} 1 \\ x \\ \Gamma_1 x \end{bmatrix} \begin{bmatrix} 1 \\ x \\ \Gamma_1 x \end{bmatrix}^T \begin{bmatrix} W_{j_1} \\ W_{j_2} \\ W_{j_3} \end{bmatrix}^T
$$

$$
= \sum_{j=1}^{m} r_j^{-1} \begin{bmatrix} W_{j_1} & W_{j_2} & W_{j_3} \end{bmatrix} \begin{bmatrix} 1 & x^T & x^T \Gamma_1^T \\ x & xx^T & xx^T \Gamma_1^T \\ \Gamma_1 x & \Gamma_1 xx^T & \Gamma_1 xx^T \Gamma_1^T \end{bmatrix} \begin{bmatrix} W_{j_1} \\ W_{j_2} \\ W_{j_3} \end{bmatrix}^T
$$

$$
= \sum_{j=1}^{m} \Big(W_{j_1} W_{j_1}^T + W_{j_2} x W_{j_1}^T + W_{j_3} \Gamma_1 x W_{j_1}^T + W_{j_1} x^T W_{j_2}^T +
$$

$$
W_{j_2} x x^T W_{j_2}^T + W_{j_3} \Gamma_1 x x^T W_{j_2}^T + W_{j_1} x^T \Gamma_1^T W_{j_3}^T +
$$

$$
W_{j_2} x x^T \Gamma_1^T W_{j_2}^T + W_{j_3} \Gamma_1 x x^T \Gamma_1^T W_{j_3}^T \Big) r_j^{-1},
$$

$$
\Omega_3 = (P_4 + P_5 \Gamma_1) \Omega_2 (P_4 + \Gamma_1^T P_5^T),
$$

$$
\Omega_4 = (P_4 + P_5 \Gamma_1) \Omega_2 (P_5 + \Gamma_1^T P_6),
$$

$$
\Omega_5 = (P_5^T + P_6 \Gamma_1) \Omega_2 (P_5 + \Gamma_1^T P_6),
$$

$$
\Omega_6 = (P_4 + P_5 \Gamma_1) \Omega_2 O(x) P_5^T + P_5 O(x) \Omega_2 (P_4 + \Gamma_1^T P_5^T)
$$

$$
+ P_5 O(x^2) P_5^T,
$$

$$
\Omega_7 = (P_4 + P_5 \Gamma_1) \Omega_2 O(x) P_6 + P_5 O(x) \Omega_2 (P_5 + \Gamma_1^T P_6)
$$

$$
+ P_5 O(x^2) P_6,
$$

$$
\Omega_8 = (P_5^T + P_6 \Gamma_1) \Omega_2 O(x) P_6 + P_6 O(x) \Omega_2 (P_5 + \Gamma_1^T P_6)
$$

$$
+ P_6 O(x^2) P_6.
$$

Moreover, $\bar{\Omega}_3$ to $\bar{\Omega}_3$, are their analogous blocks after modifying the bases, where they can also be rewritten in terms of the new variables defined in (A.4) as follows:

$$\bar{\Omega}_3 = (P_4 + \Gamma_1^T P_5^T + P_5 \Gamma_1 + \Gamma_1^T P_6 \Gamma_1)\Omega_2$$
$$(P_4 + \Gamma_1^T P_5^T + P_5 \Gamma_1 + \Gamma_1^T P_6 \Gamma_1)^T$$
$$= Z_1 W_{j_1} W_{j_1}^T Z_1^T,$$
$$\bar{\Omega}_4 = (P_4^T + P_5 \Gamma_1 + \Gamma_1^T P_5^T + \Gamma_1^T P_6 \Gamma_1)\Omega_2 (P_5 + \Gamma_1^T P_6)$$
$$= Z_1 W_{j_1} W_{j_1}^T Z_2,$$
$$\bar{\Omega}_5 = \Omega_5 = Z_2^T W_{j_1} W_{j_1}^T Z_2,$$
$$\bar{\Omega}_6 = \Omega_6 + \Gamma_1^T \Omega_7^T + \Omega_7 \Gamma_1 + \Gamma_1^T \Omega_8 \Gamma_1$$
$$= (P_4 + P_5 \Gamma_1 + \Gamma_1^T P_5^T + \Gamma_1^T P_6 \Gamma_1)\Omega_2 O(x)P_5^T$$
$$+ (P_5 + \Gamma_1^T P_6)O(x)\Omega_2 (P_4 + \Gamma_1^T P_5^T)$$
$$+ (P_4 + P_5 \Gamma_1 + \Gamma_1^T P_5^T + \Gamma_1^T P_6 \Gamma_1)\Omega_2 O(x)P_6 \Gamma_1$$
$$+ (P_5 + \Gamma_1^T P_6)O(x)\Omega_2 (P_5 + \Gamma_1^T P_6)\Gamma_1$$
$$+ (P_5 + \Gamma_1^T P_6)O(x^2)P_5^T + (P_5 + \Gamma_1^T P_6)O(x^2)P_6 \Gamma_1$$
$$= (P_4 + P_5 \Gamma_1 + \Gamma_1^T P_5^T + \Gamma_1^T P_6 \Gamma_1)\Omega_2 O(x)(P_5^T + P_6 \Gamma_1)$$
$$+ (P_5 + \Gamma_1^T P_6)O(x)\Omega_2 (P_4 + \Gamma_1^T P_5^T + P_5 \Gamma_1 + \Gamma_1^T P_6 \Gamma_1)$$
$$+ (P_5 + \Gamma_1^T P_6)O(x^2)(P_5^T + P_6 \Gamma_1)$$
$$= Z_1 \Omega_2 O(x)Z_2^T + Z_2 O(x)\Omega_2 Z_1^T + Z_2 O(x^2)Z_2^T,$$
$$\bar{\Omega}_7 = \Omega_7 + \Gamma_1^T \Omega_8 = Z_1 \Omega_2 O(x)Z_3 + Z_2 O(x)\Omega_2 Z_2$$
$$+ Z_2 O(x^2)Z_3,$$
$$\bar{\Omega}_8 = \Omega_8 = Z_2^T \Omega_2 O(x)Z_3 + Z_3 O(x)\Omega_2 Z_2 + Z_3 O(x^2)Z_3,$$

where we also used the fact that the constant term will dominate in Ω_2 as $x \to 0$. Hence, we get $\Omega_2 \to \sum_{j=1}^{m} r_j^{-1} W_{j_1} W_{j_1}^T$.

By substituting (A.3), (A.5), and (A.6) into (A.2), one can obtain

$$\dot{Z}_1 = Q + Z_1(W_2 + W_3 \Gamma_1) + (W_2^T + \Gamma_1^T W_3^T)Z_1^T$$
$$- Z_1 \left(\sum_{j=1}^{m} W_{j_1} r_j^{-1} W_{j_1}^T \right) Z_1^T + Z_2 O(x)(W_2 + W_3 \Gamma_1)$$
$$- \gamma Z_1 + (W_2 + W_3 \Gamma_1)^T O(x)Z_2^T - Z_1 \Omega_2 O(x)Z_2^T$$
$$- Z_2 O(x)\Omega_2 Z_1^T - Z_2 O(x^2)Z_2^T$$
$$= Q + Z_1 \left(W_2 + W_3 \Gamma_1 - \frac{\gamma}{2}I \right) + \left(W_2^T + \Gamma_1^T W_3^T - \frac{\gamma}{2}I \right) Z_1^T$$
$$- Z_1 \left(\sum_{j=1}^{m} W_{j_1} r_j^{-1} W_{j_1}^T \right) Z_1^T + O(x), \tag{A.7}$$

$$\dot{Z}_2 = Z_1 W_3 + (W_2 + W_3\Gamma_1)^T Z_2 - Z_1 \left(\sum_{j=1}^{m} W_{j_1} r_j^{-1} W_{j_1}^T \right) Z_2$$
$$- \gamma Z_2 + Z_2 O(x) W_3 + (W_2 + W_3\Gamma_1)^T O(x) Z_3^T$$
$$- Z_1 \Omega_2 O(x) Z_3 - Z_2 O(x) \Omega_2 Z_2$$
$$- Z_2 O(x^2) Z_3, \tag{A.8}$$

$$\dot{Z}_3 = Z_2^T W_3 + W_3^T Z_2 - Z_2 \left(\sum_{j=1}^{m} W_{j_1} r_j^{-1} W_{j_1}^T \right) Z_2$$
$$- \gamma Z_3 + Z_3 O(x) W_3 + W_3^T O(x) Z_3^T - Z_2^T \Omega_2 O(x) Z_3$$
$$- Z_3 O(x) \Omega_2 Z_2 - Z_3 O(x^2) Z_3. \tag{A.9}$$

Moreover, for the optimal control, (5.11) takes the following form:

$$u_j^* = - \begin{bmatrix} 1 \\ x \\ \Gamma_1 x + O(x^2) \end{bmatrix}^T r_j^{-1} \begin{bmatrix} 0 & 0 & 0 \\ 0 & P_4 & P_5 \\ 0 & P_5^T & P_6 \end{bmatrix} \begin{bmatrix} 0 \\ I \\ \Gamma_1 + O(x) \end{bmatrix}$$
$$\begin{bmatrix} W_{j_1} & W_{j_2} & W_{j_3} \end{bmatrix} \begin{bmatrix} 1 \\ x \\ \Gamma_1 x + O(x^2) \end{bmatrix}$$

$$= - \begin{bmatrix} 1 \\ x \\ \Gamma_1 x + O(x^2) \end{bmatrix}^T r_j^{-1} \begin{bmatrix} 0 \\ P_4 + P_5\Gamma_1 \\ P_5^T + P_6\Gamma_1 \end{bmatrix} \begin{bmatrix} W_{j_1} & W_{j_2} & W_{j_3} \end{bmatrix} \begin{bmatrix} 1 \\ x \\ \Gamma_1 x + O(x^2) \end{bmatrix}$$
$$= -r_j^{-1} x^T (P_4 + P_5\Gamma_1 + \Gamma_1^T P_5^T + \Gamma_1^T P_6\Gamma_1)(W_{j_1} + W_{j_2} x + W_{j_1}\Gamma_1 x)$$
$$= -r_j^{-1} x^T (P_4 + P_5\Gamma_1 + \Gamma_1^T P_5^T + \Gamma_1^T P_6\Gamma_1) W_{j_1}$$
$$\quad - r_j^{-1} x^T (P_4 + P_5\Gamma_1 + \Gamma_1^T P_5^T + \Gamma_1^T P_6\Gamma_1)(W_{j_2} x + W_{j_1}\Gamma_1 x)$$
$$= -r_j^{-1} x^T Z_1 W_{j_1} + O(x^2),$$

where the linear term will dominate as $x \in \bar{D}_r$, for r sufficiently small. Hence, the control rule will take the form of (5.18).

Accordingly, among the solutions of (A.7) to (A.8), only $Z_1(t)$ takes part in the control. Hence, to guarantee the stability of the closed-loop system, $Z_1(t)$ should be stabilizing as $t \to \infty$. In fact, by looking at (A.7), it is not difficult to verify it as a differential Riccati equation for the linearized system (5.15), with an additional term that is vanishing as $x \to 0$. Therefore, we can use a variant of Lemma 5.2 to conclude that the integration of (A.7) will lead to a stabilizing controller, as long as $Q + O(x) > 0$, which can be assured by assuming $x \in \bar{D}_r$, for r sufficiently small.

Furthermore, although $Z_2(t)$ and $Z_3(t)$ do not appear in the control, we require them to remain bounded. (A.8) and (A.9) can be rewritten as follows:

$$\dot{Z}_2 = (A_{cl} - \gamma I)Z_2 + G_1(Z_1) + O(x),$$
$$\dot{Z}_3 = (-\gamma I)Z_3 + G_2(Z_2) + O(x),$$

where $A_{cl} = W_2 + W_3\Gamma_1 - r_j^{-1}Z_1 W_{j_1} W_{j_1}^T$, and functions $G_1(Z_1)$ and $G_2(Z_2)$ cam be seen as inputs to these differential equations. In fact, A_{cl} is the closed-loop system matrix of (5.15), hence, is Hurwitz. Accordingly, in addition, considering $\gamma > 0$, both autonomous dynamics are asymptotically stable with Hurwitz system matrices, $A_{cl} - \gamma I$, and $-\gamma I$. This guarantees that $Z_2(t)$ will stay bounded for the bounded solution $Z_1(t)$. In a similar way, bounded $Z_3(t)$ can be concluded by the bounded $Z_2(t)$.

A.2 Supplementary Analysis of Remark 5.5

Assuming that the closed-loop system is asymptotically stable, we have $x \to 0$. Furthermore, as $\gamma \to 0$, the right-hand side of (A.7) will converge to the right-hand side of (5.19). By a variant of Lemma 5.2, we can show that $P(t)$ will converge to the solution of the Algebraic Riccati Equation (ARE) (5.16).

Index

Model-Based Reinforcement Learning: From Data to Continuous Actions with a Python-based Toolbox,
First Edition. Milad Farsi and Jun Liu.
© 2023 The Institute of Electrical and Electronics Engineers, Inc. Published 2023 by John Wiley & Sons, Inc.

Books in the IEEE Press Series on Control Systems Theory and Applications

Series Editor: Maria Domenica Di Benedetto, University of l'Aquila, Italy

The series publishes monographs, edited volumes, and textbooks which are geared for control scientists and engineers, as well as those working in various areas of applied mathematics such as optimization, game theory, and operations.

1. *Autonomous Road Vehicle Path Planning and Tracking Control*
   Levent Güvenç, Bilin Aksun-Güvenç, Sheng Zhu, Sükrü Yaren Gelbal

2. *Embedded Control for Mobile Robotic Applications*
   Leena Vachhani, Pranjal Vyas, and Arunkumar G. K.

3. *Merging Optimization and Control in Power Systems: Physical and Cyber Restrictions in Distributed Frequency Control and Beyond*
   Feng Liu, Zhaojian Wang, Changhong Zhao, and Peng Yang

4. *Model-Based Reinforcement Learning: From Data to Continuous Actions with a Python-based Toolbox*
   Milad Farsi and Jun Liu.

Printed and bound by CPI Group (UK) Ltd, Croydon, CR0 4YY

16/04/2025

14658363-0001